John Dee's Conversations with Angels

John Dee (1527–1608/9) was a Cambridge-educated natural philosopher who served Queen Elizabeth I as court astrologer and who wrote works on many subjects including mathematics, alchemy, and astronomy. His most prolonged intellectual project, however, was conversations with angels using a crystal ball and a variety of assistants with visionary abilities. Dee's angel conversations have long puzzled scholars of early modern science and culture, who have wondered how to incorporate them within the broader contexts of early modern natural philosophy, religion, and society. Using Dee's marginal notes in library books, his manuscript diaries of the angel conversations, and a wide range of medieval and early modern treatises regarding nature and the apocalypse, Deborah Harkness argues that Dee's angel conversations represent a continuing development of his natural philosophy. The angel conversations, which included discussions of the natural world, the practice of natural philosophy, and the apocalypse, were conveyed to audiences from London to Prague, and took on new importance within these shifting philosophical, religious, and political situations. When set within these broader frameworks of Dee's intellectual interests and early modern culture, the angel conversations can be understood as an attempt to practice natural philosophy at a time when many thought that nature itself was coming to an end.

Deborah E. Harkness is an associate professor of history at the University of California, Davis. She is the recipient of the Renaissance Society of America's Nelson Prize and the History of Science Society's Derek Price Award. Professor Harkness has also received fellowships from Fulbright, the National Science Foundation, ACLS, and the Huntington Library/NEH. Her articles have been published in journals such as *Renaissance Quarterly* and *Isis*.

John Dee's Conversations with Angels

Cabala, Alchemy, and the End of Nature

DEBORAH E. HARKNESS

University of California, Davis

CAMBRIDGE
UNIVERSITY PRESS

CAMBRIDGE UNIVERSITY PRESS
Cambridge, New York, Melbourne, Madrid, Cape Town, Singapore, São Paulo

Cambridge University Press
The Edinburgh Building, Cambridge CB2 2RU, UK

Published in the United States of America by Cambridge University Press, New York

www.cambridge.org
Information on this title: www.cambridge.org/9780521622288

First published 1999
This digitally printed first paperback version 2006

A catalogue record for this publication is available from the British Library

Library of Congress Cataloguing in Publication data
Harkness, Deborah E., 1965–
John Dee's conversations with angels : Cabala, alchemy, and the end of
nature / Deborah E. Harkness.
p. cm.
Includes bibliographical references and index.
ISBN 0-521-62228-X (hardback)
1. Dee, John, 1527–1608. I. Title.
BF1598.D5H37 1999
133'.092 – dc21
                                        98-41097
                                        CIP

ISBN-13 978-0-521-62228-8 hardback
ISBN-10 0-521-62228-X hardback

ISBN-13 978-0-521-02748-9 paperback
ISBN-10 0-521-02748-9 paperback

– To My Guiding Spirits –

Olive R. Harkness

John C. Harkness

and

Betty Jo Teeter Dobbs
(1930–1994)

Contents

Illustrations

Acknowledgments

The research and writing of this book was done over several years, and many libraries, institutions, agencies, and individuals played a role in its completion. I would like to thank all who provided assistance during those years.

Financial support for the research and writing of the study was provided by the Jacob K. Javits Fellowship Program; the William Andrews Clark Memorial Library of UCLA; the University of California, Davis; the Fulbright Commission; the National Science Foundation; and Colgate University.

Many individuals generously offered their time. Sandra J. K. M. Feneley of the Centre for Medieval and Renaissance Studies, Oxford, provided me with expert bibliographic assistance and help with the intricacies of early modern astrology. Rebecca Saad of the Huntington Library's Publications Office supplied much needed materials. Jonathan Smith of Trinity College, Cambridge, shared his expertise concerning the early statutes of the college. Sandy Stout and the staff of the Office of the Dean of the College of Humanities and Social Sciences at the University of California, Riverside, helped to keep open the lines of communication between a graduate student and a very busy reader. A. V. Simcock of Oxford University's Museum of the History of Science provided information about their copy of Dee's "holy table." Alison Corbett, of the Fulbright Commission, London, was a great ally and supporter during my tenure as a graduate fellow. At the University of California, Davis, Cheri Sutton of the Graduate Division and Debbie Lyon, Eteica Spencer, Elizabeth Johnson, Karen Hairfield, and Charlotte Honeywell of the History Department all provided administrative assistance above and beyond the call of duty, responding with good grace and patience to all of my extraordinary (and frequent) requests for their help.

The archival work took place in over a dozen libraries, and the staff of each one processed many requests for books and bibliographic assistance. In the United States, I am indebted to the staffs of the William Andrews Clark Library and the Folger Shakespeare Library. In Great

Britain, I owe gratitude to the following: the staff of the North Reading Room and the Manuscript Reading Room of the British Library, London; the Reverend Canon A. E. Harvey, Mrs. E. Nixon, and Dr. Tony Trowles of the Muniment Room and Library of Westminster Abbey, London; Geoffrey Davenport, Librarian, and the staff of the Royal College of Physicians Library, London; the Archivist of Corpus Christi College, Oxford; the Librarian, the Sub-Librarian, Miss N. Aubertin-Potter, and the staff of the Codrington Library, All Souls' College, Oxford; the Cambridge University Library, especially the staff of the Rare Book Room; David McKitterick, Ronald Milne, and the staff of Trinity College Library, Cambridge; the Librarian and staff of the Library of Emmanuel College, Cambridge; and the Librarian, Assistant Librarian, Mrs. E. M. Coleman, and the staff of the Pepys Library, Magdalene College, Cambridge. The Bodleian Library of Oxford University deserves special mention, for it was there that the bulk of my extensive requests for books were made, and the staff of Duke Humphrey's Reading Room and the Upper Reading Room were gracious and helpful. Julian Roberts, the Keeper of Rare Books at the Bodleian Library, generously lent his considerable expertise to me when I located *The Book of Soyga.*

Two institutions hosted me for extended periods of time during the research stages of the project, and they both gave me an atmosphere that any scholar would envy. The William Andrews Clark Memorial Library of UCLA provided me with a room of my own while my ideas were still being formed, and I thank Thomas C. Wright and his staff for those productive days. Dr. John E. Feneley of the Centre for Medieval and Renaissance Studies, Oxford, also provided me a room of my own, first as a Fulbright Fellow and then as Junior Dean. The faculty, staff, and students of the Centre offered support, criticism, and welcome interruptions for over eighteen months, and this book would not have been possible without their generosity.

Colleagues gave me support, constructive criticism, and advice at all stages of the project; they include Katharine Anderson, Samantha Meigs, Bruce Janacek, Jennifer Selwyn, and Kathleen Whalen. My friends suffered through the process of writing and helped me keep my perspective: Pat, Brynn, and Dana Herthel; Jeffrey Baines; Laura McAlinden; and Catherine VanNevel. At Colgate University I was fortunate to have Jonathan Lyon as a research assistant who lent his superb historical sensibilities to the manuscript and ferreted out every bit of information available on a wide range of topics.

I was fortunate to have the opportunity to work with Alex Holzman and Cambridge University Press during the publication process. Alex took on Dee and his angels with enthusiasm and good humor. Helen

Wheeler steered the manuscript through its final stages, and Barbara W. Folsom was an expert copyeditor. I am grateful to them all for the efforts they took on my behalf.

My committee was a source of information, inspiration, and encouragement, and I must thank Brian Copenhaver, Maggie Osler, and Ann Blair for their contributions. Maggie and Ann both were especially generous in stepping into the project during a very difficult time. Paula Findlen encouraged me from the very beginning of my research, and it was with her guidance that my first tentative steps into the wonderful world of angels took place.

Karen Halttunen, partner and friend, cast her careful eyes over the manuscript and opened her sympathetic ears to the author in good times and bad. Without her unshakeable faith in me and her unflagging interest in my ideas, this book could not have been completed. I thank her for her love and support.

Finally, Betty Jo Teeter Dobbs profoundly influenced this book. Few students are fortunate enough to work under the supervision of so wise and generous an adviser. For six years Jo encouraged my enthusiasm for things occult, provided constructive criticism, and gave me the intellectual latitude to consider every possible idea – no matter how remote – that might help to explain Dee's interest in angels. During that time her support and enthusiasm never wavered. She was a formidable scholar, an inspiring teacher, a generous spirit, and a treasured friend; I will miss her.

It is to Jo, and to my parents (who have given me more than I can ever fully acknowledge), that this work is dedicated.

Abbreviations and Conventions

The following abbreviations are used in the notes for frequently cited works. Full information on these titles can be found in the Select Bibliography.

Agrippa, *DOP*	Henry Cornelius Agrippa, *De occulta philosophia libri tres* (1533)
Clulee	Nicholas H. Clulee, *John Dee's Natural Philosophy: Between Science and Religion*
Dee, *AWS* II	John Dee, *John Dee's Actions with Spirits*, edited with an introduction by Christopher Whitby. 2 vols. Vol. II
Dee, *CR*	John Dee, *Compendious Rehearsal* (1592), in *Autobiographical Tracts of Dr. John Dee*, edited by James Crossley
Dee, *MH*	John Dee, *Monas hieroglyphica* (1564), translated by C. H. Josten
Dee, MP	John Dee, "Mathematical Preface," in *The Elements of Geometrie of . . . Euclid of Megara* (1570), translated by Henry Billingsley
Dee, *PA*	John Dee, *Propaedeumata Aphoristica*, edited and translated by John Heilbron and Wayne Shumaker
Dee, *PD*	John Dee, *Private Diary of Dr. John Dee*, edited by J. O. Halliwell
French	Peter J. French, *John Dee*
Roberts and Watson	Julian Roberts and Andrew Watson, *John Dee's Library Catalogue*
TFR	John Dee and Edward Kelly, *A True and Faithful Relation of what happened for many years . . .*, edited by Meric Casaubon

| Thorndike | *A History of Magic and Experimental Science.* 8 vols. |
| Whitby, *AWS* I | John Dee, *John Dee's Actions with Spirits*, edited with an introduction by Christopher Whitby. 2 vols. Vol. I |

Quotations

When quoting from the angel conversations, I have retained the original spelling, punctuation, and capitalization. Any marks of emphasis that appear are Dee's own. Carets indicate text that has been inserted above the line in the manuscript. Additional letters have been supplied in square brackets when Dee's spelling might lead to a misreading of the text.

Dates

Based on the evidence from the angel diaries, Dee began each new year as we do, on the first of January. When he was on the Continent from 1583 to 1587, Dee began gradually to incorporate the new calendar into his entries, noting both "stylo novo" and "stylo veteri." For clarity of reference, the date that Dee recorded in his diary entries has been maintained, even though this may cause some chronological confusion.

Items from Dee's Library

When reference is made to one of Dee's books, a notice is made of the number assigned to it by Roberts and Watson. Bibliographic details are thereby kept as brief as possible, and readers are directed to Roberts and Watson for more information on a given title, multiple editions, present location, shelfmark, and so on. For Dee's books that contain marginalia cited in this work, full bibliographic information, including current location if known, is included in the Select Bibliography.

Introduction

> We know in part, and we prophesy in part. But when that which is perfect is come, then that which is in part shall be done away.
>   When I was a child, I spake as a child, I understood as a child, I thought as a child:
> but when I became a man, I put away childish things.
>   For now we see through a glass, darkly;
> but then face to face: now I know in part;
> but then I shall know. . . .
>
>   1 Corinthians 13:9–12

Between 1581 and 1586, and again in 1607, Elizabethan England's most highly regarded natural philosopher, John Dee, talked with angels about the natural world and its apocalyptic end. With the aid of an assistant, or "scryer," and a crystal called a "showstone," Dee attempted to see through the dark days of his own time and into what he hoped was a bright and promising future. Scattered through several manuscript collections in the Bodleian and British Libraries,[1] Dee's records of these conversations now represent one of the early modern period's most enduring intellectual mysteries: why would a Cambridge graduate who boasted the title "the Queen's philosopher" engage in such a seemingly fruitless, apparently groundless, and enormously time-consuming activity? Was Dee a gullible fool?[2] Had he suffered a mental breakdown?

---

1. Dee's angel diaries are now British Library Sloane MSS 3188, 3189, 3191; British Library Add. MS 36674 and Bodleian Library MS Ashmole 1790.
2. See Frances A. Yates, *The Theatre of the World* (Chicago: University of Chicago Press, 1969), pp. 5–18, for a sensitive but still guarded interpretation. John Heilbron, in his introduction to John Dee's *Propaedeumata aphoristica*, demonstrated less sensitivity to medieval and early modern beliefs, describing Dee's latter years as a balance of "more lucid moments," when the natural philosopher attempted to "clear himself of charges of invoking spirits," and less lucid moments spent conversing with angels. J. L. Heilbron, "Introductory Essay," in *John Dee on Astronomy: Propaedumata aphoristica*, ed. and trans. by Wayne Shumaker (Berkeley: University of

Given these serious reservations about Dee and his conversations, historians of science have wondered if the angel diaries can yield any useful information to scholars specifically interested in the practice of natural philosophy in the late sixteenth century or illuminate the cultural and intellectual world of Elizabethans more generally.[3]

The answers to these questions can be found, but not exclusively within the pages of Dee's manuscript angel diaries. The diaries offer only pieces of the puzzle, fragmentary remnants of a vast intellectual undertaking. Today, only a small number of them remain; the greater proportion were destroyed by Dee and by a zealous seventeenth-century kitchen maid who used the pages from the volumes to line her employer's pie plates.[4] But even were the diaries complete, we would still not find all of the answers we seek. Like the transcripts of court trials or the written inventory of a library, the diaries that remain are full of telling silences and ellipses, devoid of the nuances of voice and gesture that could tell us so much about what happened when Dee looked into his showstone.

We still, however, can look for further clues into Dee's mysterious project. First, we can examine the treasures of his extensive library to

California Press, 1978), p. 15. Shumaker interpreted the angel diaries as evidence that Dee was gullible to a fault, pointing out that "even in his own period few men were so susceptible to the portentous" (Wayne Shumaker, "John Dee's Conversations with Angels," in *Renaissance curiosa* 15-52 [Binghamton, NY: Center for Medieval and Early Renaissance Studies, 1982] p. 48). But he was forced to admit that Dee appeared entirely rational throughout, noting he "does not lose control, he does not rave, but transcribes the visions and conversations reported by his scryers with sober accuracy" (ibid., p. 22).

3. While Clulee pointed out the important links between Dee's studies and the natural philosophers who preceded him, he was still unable to reconcile these traditions with Dee's interest in the angel conversations. Dee's angel diaries remained an intellectual aberration to Clulee, one that "cannot be considered as science or natural philosophy," despite their inclusion of concepts from the cabalistic and alchemical traditions (Clulee, p. 203). Christopher Whitby's analysis of the earliest angel diaries, the history of crystal gazing, and biographies of the chief participants in the conversations remains the cornerstone of scholarship on these matters. See John Dee, *John Dee's Actions with Spirits*, ed. with intro. by Christopher Whitby, 2 vols. (New York: Garland Publishing, 1988), passim.

4. For Dee's description of the destruction of his manuscripts, see Dee, in C. H. Josten, ed. and trans., "An Unknown Chapter in the Life of John Dee," *Journal of the Warburg and Courtauld Institutes* 28 (1965): 223-257 hereafter cited as Dee in Josten, and Dee, PD, entries for 1607. For Ashmole's introduction see Dee, AWS II: 2-4.

see if we can piece together his intellectual rationale for the conversations. Frances Yates believed that the intellectual richness of the Renaissance was contained within Dee's library, and its contents have recently been studied and every attempt made to account for the volumes he owned and to make them accessible.[5] Not only do the authors of Dee's books offer us further insights into why he undertook to talk with angels; Dee himself left a trail for us to follow in the marginal comments he made in his books.[6] Underlined passages, references to other authors, and accounts of his own life experiences lead like a path of footprints to the angel conversations, helping to contextualize and ground them.

We can also cast our eyes beyond Dee's intellectual training and his personal experiences and try to situate his conversations with angels in the broader cultural context of the second half of the sixteenth century. This period was vital and chaotic: deepening religious divisions, sharp political disagreements between England and her European neighbors, a renewal of apocalyptic fervor, and the growth of print culture all helped to make it so. Information from newly discovered lands fostered a sense of crisis among the intellectual elites, as they tried to reconcile new worlds with old texts.[7] Cities like London struggled to accommodate the growing number of people seeking employment, anonymity, or religious refuge within their bursting walls. Those who could not be so accommodated began to cross the Atlantic. Spies and intelligencers of every stripe traveled the countryside, gathering news of religious radicalism, political conspiracy, and heresy. Dee's contemporaries were often pessimistic about what they perceived as a chaotic state of affairs and thought that the end of the world was certainly near.

The marginalia in Dee's library, when tracked through the late-sixteenth-century belief that the world was coming to an end, lead straight to his angel conversations. These conversations did not represent

5. Yates, *Theatre of the World*, pp. 1–19. Dee's final extant library catalogue was completed in 1583 just before his trip to Europe with Edward Kelly. Roberts and Watson argue persuasively that the 1583 list contains annotations indicating which books were selected for Dee's traveling library. Roberts and Watson, pp. 53–54.

6. For two different interpretations of the marginalia and their significance, see William Sherman, *John Dee: The Politics of Reading and Writing in the Renaissance* (Boston: University of Massachusetts Press, 1995), and Deborah E. Harkness, "The Scientific Reformation: John Dee and the Restitution of Nature," Ph.D. diss., University of California, Davis, 1994.

7. See Anthony Grafton, *Defenders of the Text: The Traditions of Scholarship in an Age of Science 1450–1800* (Cambridge, MA: Harvard University Press, 1991).

a break in Dee's natural philosophy or intellectual development; instead, they furthered and supported Dee's interest in the natural world while taking into account the particular challenges presented by the times in which he lived. In particular, the angel conversations confirmed Dee's belief that the natural world was analogous to a text. But the Book of Nature was not a reliable text; it was an imperfect, corrupt, and decaying text that could not be read properly. The angels gave Dee the exegetical and restorative tools to read, understand, and rectify the Book of Nature. The angel conversations thus represent Dee's attempt to practice natural philosophy at a time when many thought that nature, time, and the world as they knew it were approaching their end.

The angel conversations reveal Dee's belief that a key aspect of his role as a natural philosopher practicing his art at the end of nature was to communicate – with angels, with patrons, and the public – information about the Day of Judgment that was near at hand. The first chapter describes the conversational and communicative qualities of Dee's enterprise, shifting attention away from the angel *diaries* as static texts and focusing instead on the personalities and properties that were involved in each conversation. Chapter 1 also examines the ways in which Dee's angel conversations were perceived by the people who took part in them, from the king of Poland to scryers. Few of the individuals who experienced the conversations expressed doubts about them, though some questioned Dee's ability to serve as an apocalyptic prophet.

Chapter 2 considers the genesis of the angel conversations. Drawing on information from Dee's library as well as his own published natural philosophical works, this chapter argues that for some time Dee had been looking for a way to close the gap between the perfect, immutable heavenly spheres and the corrupted Book of Nature. Manuscript marginalia in his books reveal his search for a universal science that would be capable of understanding, perfectly and with certainty, the mysterious workings of the natural world. In each of Dee's works we can see his struggle to extend a ladder from the deteriorating world to the heavens.

As Chapter 3 explains, Dee's frustration over his failure to construct a universal science, given the natural philosophical methods available, led him to a new course of action: the angel conversations. Dee's marginalia are particularly important to our understanding of how he arrived at this decision, for he left no single synthetic analysis of angels and their place in the Book of Nature. Scattered notes, underlinings, and marks of emphasis must instead tell us what Dee found most compelling about angels in the many books he owned and studied. These annotations clarify his belief that angels comprised the perfect intermediary between God and humans, celestial and terrestrial, sacred and mundane. Intermediary agencies had long fascinated Dee – he discussed the intermediary potential of everything from mathematics to hieroglyphics to light in his published works – but it was not until he constructed an optically grounded method for communicating with angels that his efforts generated satisfactory results.

Dee's excitement about the angel conversations becomes understandable only when seen through the lens of Reformation culture. Although England had embarked on its own religious path when Dee was a child, common beliefs fueled most sixteenth-century reform movements, both Protestant and Catholic, as Chapter 4 outlines. A belief in eschatological signs, for example, fostered a sense that the current human condition was nearing its end and would soon be replaced by the New Jerusalem, where people would live in peaceful harmony with a perfect knowledge of the world and its mysteries. From accounts of rains of blood to monstrous births, the early modern literate public had an insatiable interest in cataloguing and contesting signs of the end of nature appearing around them. One eagerly anticipated eschatalogical sign was increased communication between the celestial and terrestrial levels of the cosmos. Angelic messages had preceded many events of cosmic importance, such as the birth of Christ, and angels were harbingers of the end of days in the prophetic biblical revelations of Saint John. Before the Fall, Adam had enjoyed communication with God and the angels in Paradise. Dee's conversations with the angels thus conformed to a contemporary sense that God, through his intermediary angels, would alert some select individuals to imminent events.

Dee still had to grapple with the problem of interpreting the decaying Book of Nature, however. The angels revealed an interpretive tool that would further his efforts: the "true cabala of nature." Chapter 5 describes this angelic cabala and explains how it is similar, and dissimilar, to the Jewish and Christian cabala of the early modern period. Like many other aspects of Dee's angel conversations, the cabala of nature was not entirely compatible with existing early modern systems of thought. Such incompatibilities tell us a great deal about the methodological limitations facing a natural philosopher like Dee as well as his aspirations to apply natural philosophical techniques to some of the most pressing problems facing society.

After interpreting the Book of Nature, Dee was expected to heal the corrupted text by using an angelically revealed "medicine." In Chapter 6 the angels' medicine is put in the context of Dee's alchemical studies and the alchemical beliefs of his contemporaries. Alchemy was one of the most promising, as well as most frustrating, early modern natural philosophical enterprises. Few alchemists dared to boast of success in transmuting base metals into gold, and even fewer believed that they had achieved the highest goal of all: the transmutation of the human body from mortal and corruptible flesh to incorruptible quintessence. Dee and some of his contemporaries thought that their failures were due in part to the flawed methods they employed; others, that their own imperfections hampered their efforts. Dee hoped, through the information imparted to him by his angelic "schoolmasters," to practice a restored alchemy that would finally yield positive results.

With the Book of Scripture in hand and the Book of Nature before him, Dee was attempting to refashion the identity of the natural philosopher to include a reinterpretation of knowledge, a universal reform of institutions, and a restitution of nature and all things. In his angel conversations, the Old Testament prophet receiving revelations from God merged with the New Testament magus responsible for interpreting nature. This combination provided a new direction for natural philosophy that eventually shaped the aspirations of many seventeenth-century natural philosophers. At the same time, however, Dee's approach remained faithful to the ideas of the past. Just as his angel conversations served as a liminal exchange between celestial and terrestrial, so Dee emerges from this study as a liminal figure between medieval and modern, magical and scientific, Protestant and Catholic.

PART I

Genesis

And certainly He to whom the whole Course of Nature lyes open, rejoyceth not so much that he can make Gold and Silver, or the Divells to become Subject to him, as that he sees the Heavens open, the Angells of God Ascending and Descending, and that his own name is fairely written in the Book of Life.

—Elias Ashmole,
"Prologomena" to the *Theatrum Chemicum Brittanicum*
(1652)

1

The Colloquium of Angels

Prague, 1586

A learned and renowned Englishman whose name was Dr. De[e]: came to Prague to see the Emperor Rudolf II and was at first well-received by him; he predicted that a miraculous reformation would presently come about in the Christian world and would prove the ruin not only of the city of Constantinople but of Rome also. These predictions he did not cease to spread among the populace.

—Lutheran Budovec,
Circulo horologi lunaris[1]

Lutheran Budovec lived in Prague in the latter half of the sixteenth century, and when he noted in his journal the activities of John Dee he profiled a man far different from the "magus of Mortlake" with whom we are familiar.[2] An ambassador, a popular prophet, and a religious rebel – all are suggested in Budovec's brief remarks, but the natural philosophical persona we have come to associate with Dee is absent. The link between Dee's personae, as outlined by Budovec, is *communication*, a side of Dee's life and activities explored only recently but which is emerging as an important feature of his enigmatic intellect. Recently described as an "arch-communicator" of ideas, Dee is beginning to be seen as both a contemplative natural philosopher and a vocal participant in the intellectual and cultural life of late-sixteenth-century Europe.[3]

1. Quoted in R. J. W. Evans, *Rudolf II and His World* (Oxford: Clarendon Press, 1973), p. 224.
2. The most reliable and comprehensive recent biographies of Dee are Peter French, *John Dee: The World of an Elizabethan Magus* (London: Routledge and Kegan Paul, 1972), and Nicholas Clulee, *John Dee's Natural Philosophy: Between Science and Religion* (London: Routledge, 1988). See also Charlotte Fell Smith, *John Dee (1527–1608)* (London: Constable and Company, 1909).
3. See William H. Sherman, *John Dee: The Politics of Reading and Writing in the English Renaissance* (Amherst and Boston: University of Massachusetts Press, 1995).

Such a combination of activity and contemplation can be seen vividly in the records Dee kept of his conversations with angels as well as the events that surrounded them. Throughout, we see Dee the "arch-communicator" in a new light: as a natural philosopher prepared to engage in conversation with all levels of the cosmic system in an effort to come to terms with the intricacies of natural philosophy and the state of the natural world.

On 10 April 1586, John Dee made an account of the hum of activities in his household, which had temporarily relocated to the central European city of Prague. It was a time of crisis, when Dee believed a "great catastrophe" was "overhanging the world." His life and career were at a critical juncture.[4] His local patron, the Holy Roman emperor Rudolf II, was angry and frustrated. Though Dee had arrived in the city nearly two years earlier, he had not been as useful in the emperor's alchemical experiments as Rudolph expected. In Rome, the center of European Catholicism, word of Dee's prophecies about a "miraculous reformation" reached the ears of the pope, who became so concerned that officials launched an inquiry into Dee and his activities. Dee's assistant, Edward Kelly, was threatening to part company and was, in any event, a difficult, troublesome man. Kelly, unlike Dee, was absorbed with the emperor's alchemical problems and wanted more time to devote to them. He had, in addition, started to confess his myriad sins to the nearby Jesuits, thus increasing Catholic awareness of Dee and his household. As if this were not enough to worry England's foremost natural philosopher, Dee's wife was annoyed with him too. Their family and household had been on the road since the autumn of 1583, their children had to be fed and cared for, and she detested Edward Kelly and his disruptive presence.[5] But although the crisis facing Dee varied in its manifestations, the root of all the problems could be found in a single place: his conversations with angels.

The crisis of April 1586 marks the beginning of the end of Dee's best-documented efforts to communicate with angels. Though this seems an odd moment at which to enter into the challenging mental and cultural

4. These remarks, and a single conversation also dated 10 April 1586, were separated from the other diaries that contain Dee's angel conversations. Now Bodleian Library Ashmole MS 1790, ff. 1–10, they can be found in "Dee in Josten." Dee in Josten, p. 226.

5. For a detailed description of the Dee household and the important role of Jane Dee in its management, see Deborah E. Harkness, "Managing an Experimental Household: The Dees of Mortlake and the Practice of Natural Philosophy," *Isis* 88 (1997): 242–262.

world of Dee and his contemporaries, it is often this later image of the angel conversations which makes the deepest impression: the slow deterioration of an intellectual project with implications too vast for its audience to comprehend, too difficult for its participants to sustain. Interpersonal conflict, suspicion, patronage difficulties, notoriety, and a confusion of purpose have come to characterize what most scholars dismiss as evidence of John Dee at his most gullible, of Edward Kelly at his most deceptive, of Jane Dee at her most exasperated, and of Rudolf II at his most skeptical. Such a characterization, as the next chapter will discuss, is inaccurate and misleading. Still, it is useful to begin the angel conversations when they were at their most volatile and complicated, when all that Dee was struggling to keep coherent was about to dissolve into chaos.

Orienting oneself in the complex, multivalenced, and densely populated world of John Dee is not easy, and no one moment in the conversations can serve as a guide to the entire body of evidence now at our disposal. The solitary magus that so many conjure up when thinking of Dee is particularly difficult to find in the angel diaries, with their records of conversations with angels, court officials, alchemical assistants, papal nuncios, Jesuits, and household members. This chapter serves as a general introduction to Dee's complicated world – the people, angels, and properties that appear throughout his angel conversations. With this information, we will be in a position to retrace our steps and go back to the earliest surviving conversations from this more familiar juncture in the life of John Dee.

Throughout the chapter, Dee's interaction with the angels will be referred to as *conversations*. Though some scholars refer to the conversations as "spirit actions," and others focus on the "angel diaries" that record them, Dee himself described his experiences as the "colloquium of angels." To Dee the conversations were a group discussion, an ongoing attempt to reconcile what was known about the natural world with the unknown and mysterious. Emphasizing the conversational nature of the events makes their dynamism more apparent. Like most conversations, Dee's angel conversations can abruptly stop in the middle of an important exchange or contain allusions to people and places outside the parameters of the recorded conversation. Dee was not intent merely on contacting angels, nor in keeping detailed diaries of those conversations. He was determined, as his own designation implies, to communicate and discuss his concerns about the natural world and its future with the angels, his associates in the conversations, and, as Budovec reminds us, with a broad popular audience.

Dee's audiences for the conversations would have been receptive to his visionary experiences – at least in theory. But the tangled policies of the Reformation and Counter-Reformation made authentication of such experiences difficult and dangerous. In Dee's case, his travels through Europe compounded both the difficulties and the dangers because the angel conversations had to be authenticated by Protestants and Catholics with divergent beliefs concerning miracles, the relevance of private revelations to public reform, and the extent to which people outside of church control could and should enjoy visionary activity. Nonetheless, Dee divulged the contents of his conversations to "the worthy, namely to those . . . pious, humble, modest, sincere, [and] conspicuous in Christian charity."[6] Dee's faith in his angelic guides and their messages was unshakeable, and he believed that whenever his conversations were "recited in public, every later age will rejoice with unbelievable exultation, because the strength of divine truth, wisdom, and power is invincible."[7] Comprehending the breadth and complexity of Dee's "colloquium of angels" and the roles of its many participants is crucial to our understanding the content of the revelations.

John Dee, Natural Philosopher

The most important single person who took part in the angel conversations was John Dee. Because the accidents of history have helped to make Dee into a lonely and somewhat idiosyncratic figure wedged historically between the medieval genius of Roger Bacon and the modern foresight of Francis Bacon, we are no longer steeped in the intellectual traditions that shaped him and find ourselves at a loss to understand what he took for granted about the natural world and his place in it. In his own time, however, Dee was one of England's most sought-after scholars, who held valued opinions on a wide range of topics. For example, he was offered a position as a lecturer at Oxford University, for example, which he declined. He was consulted about the most astrologically auspicious day for Elizabeth I's coronation. He entertained guests including Sir Walter Raleigh, the earl of Leicester, and the Dutch geographer Abraham Ortelius at his home outside of London on the river Thames. At Mortlake, Dee was generous with his expertise and with the unparalleled resources of his famous library. For visitors unable to make the journey to Mortlake, Dee would interrupt his work to go to them, as he did on 23 February 1581, when the French ambassador,

6. Dee in Josten, p. 226.   7. Ibid., Josten, p. 227.

Michel de Castelnau, introduced him to Jean Bodin in the royal Presence Chamber at Westminster.[8]

The relationships Dee had with the intellectual and political elite of his own country were matched, if not exceeded, by those he enjoyed on the European continent. He made his mark there while still in his twenties, lecturing at Paris, studying at the University of Louvain, and meeting fellow scholars and potential patrons in Italy, the Low Countries, and the Holy Roman Empire. Peter Ramus, the famous French humanist and methodologist, considered Dee one of two learned men in England.[9] The Italian mathematician Federico Commandino described Dee to his patron, Francesco Maria II of Urbino, as "a man of excellent wit, and singular learning."[10] Even late in his life Dee was not entirely forgotten by his colleagues abroad but was in communication with natural philosophers like Tycho Brahe, who sent him a copy of *De mundi aetherei recentioribus phaenomenis* (1588).[11]

The heights that Dee reached were somewhat unusual – though not completely unique – for a man born into the gentry during a period when upward mobility was enhanced by education and the acquisition of minor positions at court or in great households. Born on 13 July 1527 to Rowland Dee, a Welsh gentleman server to Henry VIII, and his English wife, Jane (sometimes Joanna) Wild, young John Dee was educated as carefully and thoroughly as his parents could afford. He was sent to Chelmsford Grammar School, Essex, around 1535 and became "well-furnished with understanding of the Latin tongue."[12] Furnished

---

8. Dee, *PD*, p. 10. For details of Dee's life and his position in the English court and academic circles, readers are directed to Clulee, especially pp. 19–29; French, *John Dee*; and Deborah E. Harkness, "The Scientific Reformation: John Dee and the Restitution of Nature" (Ph.D. diss., University of California, Davis, 1994), pp. 75–144. For further information on Bodin and his place in the history of science, see Ann Blair, *The Theater of Nature: Jean Bodin and Renaissance Science* (Princeton, NJ: Princeton University Press, 1997).

9. J. A. Van Dorsten, *The Radical Arts: First Decade of an Elizabethan Renaissance* (London: Oxford University Press, 1970), pp. 16–20.

10. Letter from Commandino to Francesco Maria II, quoted in J. L. Heilbron, "Introductory Essay," in *John Dee on Astronomy: Propaedumata aphoristica*, ed. and trans. by Wayne Shumaker (Berkeley: University of California Press, 1978), p. 16.

11. See Roberts and Watson, #D5.

12. Dee recounts many details about his early life in his *Compendious Rehearsal*. For his early education, see Dee, *CR*, pp. 4–5.

with the essential linguistic foundation on which all subsequent humanistic scholarship would rest, Dee was well prepared to enter St. John's
College, Cambridge, in November 1542 at the age of fifteen. His family
provided all the funds necessary for Dee to satisfy his intellectual and
personal needs, including his desire to begin collecting a scholarly library.

In the 1540s St. John's College was a vibrant, intellectually stimulating
atmosphere.[13] Distinguished alumni and instructors included Bishop
John Fisher, Edward VI's tutor John Cheke, Elizabeth I's tutor Roger
Ascham, and the queen's future Lord Treasurer and Secretary of State,
William Cecil. Dee's curriculum, for the most part, was based on the
traditional subjects (Aristotle, ancient languages, rhetoric, and some
mathematics), although lectures on the Scriptures in both Hebrew and
Greek, music, higher mathematics, and philosophy were also available.[14]
Dee was awarded the bachelor of arts degree early in 1546, and was
appointed a fellow of the college shortly thereafter. By the end of the
year, however, he was elected as one of the original fellows and under-
reader of Greek at Trinity College, Cambridge University's newest foundation.[15] At Trinity, Dee entered a more mature, and more adventurous,
stage in his intellectual and professional life.

During the summer vacation of 1547, before taking up his position at
Trinity, Dee first traveled to Europe. According to him, the purpose of
the visit was "to speak and confer with some learned men . . . chiefly
Mathematicians." The individuals Dee mentions – Gemma Frisius, Gerard Mercator, Johannes Gaspar á Mirica, and Antonius Gogava – all

13. For more on Cambridge University and the St. John's College curriculum
    in this period, see *Early Statutes of the College of St. John the Evangelist
    in the University of Cambridge*, ed. J. E. B. Mayor (Cambridge: Cambridge
    University Press, 1859); David McKitterick, "Two Sixteenth-Century Cat-
    alogues of St. John's College Library," *Transactions of the Cambridge
    Bibliographical Society* 7 (1978): 135–155, p. 138; Clulee, pp. 22–26.

14. This combination of studies and resources may have given Dee the oppor-
    tunity to explore some of the natural philosophical and occult studies upon
    which he would later focus. See Mordechai Feingold, "The Occult Tradi-
    tion in the English Universities of the Renaissance," in *Occult and Scientific
    Mentalities in the Renaissance*, ed. Brian Vickers (New York: Cambridge
    University Press, 1984); pp. 73–94.

15. For more information on Trinity, see G. M. Trevelyan, *Trinity College
    (1943)* (Cambridge: Trinity College, 1972); Damien Riehl Leader, *A His-
    tory of Cambridge University: The University to 1546*, vol. 1 (Cambridge:
    Cambridge University Press, 1988), especially pp. 345–346.

were associated with the University of Louvain, which was gaining a reputation for natural philosophical and occult studies.[16] This visit made such an impression on Dee that he returned to Louvain in 1548 after he received his master's degree from Trinity, visiting Antwerp, Brussels, and Paris. During his 1548 journey he was introduced to intellectuals including Abraham Ortelius, Peter Ramus, and Jean Fernel.[17]

Trips to the Continent and associations with foreign natural philosophers and scholars became a common feature of Dee's intellectual life, and it is clear that he influenced, and was influenced by, an extraordinarily rich and varied group of scholars, navigators, politicians, and patrons. His working habits mirrored this cosmopolitan profile. Far from being a solitary figure, Dee rarely worked in isolation; a number of his intellectual projects were undertaken in partnership with other scholars. With Federico Commandino, for example, he prepared an edition of *De superficierum divisionibus*, a work by Euclid wrongly attributed to Machomet Bagdedine.[18] With Henry Billingsley, Dee worked on the first English edition of Euclid's *Elements*.[19] His alchemical experiments required assistants, and his most notorious scryer, Edward Kelly, served as his associate not only while communicating with angels but also in the alchemical laboratory.[20] Dee's *Monas hieroglyphica* was prepared in the house of a Dutch printer, Willem Silvius, with whom Dee lived from 1562 to 1564.[21] These intellectual connections support William Sherman's contention that Dee was not a solitary magus engaged primarily in contemplation but an active member of European intellectual circles.[22]

Part of Dee's communicative impulse extended to posterity, and the difficult events surrounding 10 April 1586 precipitated a typical re-

---

16. Dee, *CR*, pp. 5–6. For the University of Louvain's connections to scientific interests, see H. A. M. Snelder, "Science in the Low Countries during the Sixteenth Century," *Janus* 70 (1983): 213–227, p. 214. For more on these individuals and their relationship to Dee, see Harkness, "The Scientific Reformation," pp. 94–108.

17. See Harkness, "The Scientific Reformation," pp. 108–118.

18. Further information on this collaboration can be found in Enrico I. Rambaldi, "John Dee and Federico Commandino: An English and an Italian Interpretation of Euclid during the Renaissance," *Rivista di Storia della Filosofia* 44 (1989): 211–247.

19. Dee, "Mathematical Preface," in Euclid's *Elements* (London: 1570).

20. See, for example, Dee, *PD*, p. 24.

21. For more on this period of Dee's life, see Harkness, The Scientific Reformation," pp. 133–135.

22. Sherman, *John Dee*, pp. 8–17.

sponse: he wanted to record how and why he communicated with angels so that his contemporaries and future generations would not misunderstand.[23] Despite his utter faith in the authenticity of the angel conversations, Dee was sensitive to the ways in which they might be perceived by the surprisingly large number of people who knew of them. In a document intended for a readership beyond his immediate circle, Dee listed some of the charges that could, and indeed had, been leveled against him and his associates. Should "somebody ... dare to assert that we are most subtle impostors," Dee wrote, or that the angel conversations

> were fabricated by us and that for that reason we should be deemed pernicious citizens in the Christian polity ... or that we are over-credulous or doting fools ... drawn and driven into error ... by some very astute evil spirit ... or if somebody should wish to maintain ... that, in our time and in the present condition of the world, all revelation by divine communication has ceased,

then Dee was prepared to offer compelling evidence to support his angel conversations.[24]

Dee and His Scryers

Dee conversed with angels about a variety of topics ranging from natural philosophy and politics to the end of the world and the best treatment for his wife's illnesses. Each of Dee's angel conversations involved an assistant, technically referred to as a "scryer." Scryers traveled about in early modern England, making a living through their reputation for visions and trying to maintain a balance between the economic necessity of adopting a public persona as a practicing seer and an interest in maintaining secrecy for purposes of self-preservation.[25] Dee obtained the services of at least four scryers: Barnabas Saul, Edward Kelly, his own son Arthur Dee, and Bartholomew Hickman. In addition, it is highly probable that Dee employed at least one additional scryer prior to 1581.[26] Typically, Dee met his scryers through the agency of patrons,

23. These remarks, and a single conversation also dated 10 April 1586, were separated from the other collections of Dee's angel conversations. Now Bodleian Library Ashmole MS 1790, ff. 1–10. See Dee in Josten, pp. 223–257.

24. Dee in Josten, p. 227.

25. Keith Thomas, *Religion and the Decline of Magic* (New York: Charles Scribner's Sons, 1971), pp. 215 and 230.

26. See Whitby, *AWS* I: 52–54 for his analysis of possible scryers, including William Emery, Roger Cook, and Robert Gardner. Scholarly consensus is

scholars who used his library, and people intimate enough to know of his interest in contacting angels. Bartholomew Hickman, for example, was introduced to Dee in 1579 by one of Queen Elizabeth I's favorites, Sir Christopher Hatton. Dee and Hatton met at Windsor in 1578 when Hatton was knighted, and the two men struck up a friendship that led to Hickman's introduction to Dee.[27] From Barnabas Saul to Edward Kelly and on to Bartholomew Hickman, Dee maintained a relationship with a scryer for the latter third of his life with no significant break in the quality or quantity of the angel conversations.

Scrying was already an ancient method of divination in the sixteenth century, for the practice of looking into a shiny or reflective object to aid in prophecy had a long history extend back to the Greeks.[28] At the same time, this form of divination is not consistent with later spiritualistic practices, and Dee's scryers were not mediumistic in the Victorian sense. Spirits of the dead did not speak through them, and the angel conversations cannot properly be called séances.[29] Instances in the angel conversations when the scryer heard a voice without consulting the showstone were uncommon, and Dee's use of a specific physical object to facilitate visions was part of the tradition. Scryers were the skilled laborers who made such an inquiry into the mysteries of nature possible.

Barnabas Saul is the first scryer who can be positively identified from the surviving records. He took part in the first extant angel conversation, which occurred on 22 December 1581 (Figure 1). The conversation's format in the diary closely resembles all subsequent conversations, regardless of the specific scryer involved. Dee first sets the scene in a descriptive note, marking his role in the conversations with his personal hieroglyph, the delta. The angel's responses are marked by a two-letter abbreviation of its name – here "An" for the angel Anael. Dee continued his practice of making marginal notes in his library books in the angel diaries, with asides, later notes, illustrations, and additional remarks all

that William Emery was Dee's earlier scryer. See Clulee, pp. 140–141, and Roberts and Watson, p. 47.

27. Dee, *PD*, p. 5.

28. See Theodore Besterman, *Crystalgazing: A Study in the History, Distribution, Theory, and Practice of Skrying* (London, 1924); Thomas, *Religion and the Decline of Magic*, p. 230; Whitby, *AWS* I:75–93; Armand Delatte, *La Catoptromancie grecque et ses dérivés*, Bibliothèque de la Faculté de Philosophie et Lettres de l'Université de Liège, vol. 48 (Paris/Liège, 1932).

29. The practice of scrying differs markedly from later forms of spiritualism, especially Victorian trance mediumship. See Alex Owen, *The Darkened Room: Women, Power, and Spiritualism in Late Victorian England* (Philadelphia: University of Pennsylvania Press, 1990).

Anno 1581. Decembris 22. Mane.   (3)   Mortlak

ANAEL

Δ   After my fervent prayer made to God, for his mercifull
    comfort and instruction, &c. ... in the name of God
    &c. and the good Angel, named Anael ... that he
    his Divine power &c) I willed the Skryer (named
    Saul) to loke into my great Chrystalline Globe, If
    God had sent his holy Angel Anael, or no.

⊙   Saul loking into my ... Stone (or chrystall globe
    for to espye Anael, &c. Saul e ... one, wepyng ...
    ... being ... requested of me to tell
    the Truthe, if he were Anael, He ... did appere
    very bewtifull, with apparell yellow glittering like gold: and his
    head had beames like star beames, blasing, and spredding from it; his eyes
    fyrie.   He wrote (which he ... in the stone here) in ... letters, and the
    letters seemed of transparent gold. which, Saul was not able eyther
    presently to reade, that I myght write after his voyce, neyther to
    imitate the letters in short tyme.

A bryght starr did go up ... and downe by him.
There appeared also a white doe, with a long tail.
And many other visions appeared, with this second: the first being vanished
quite away.   Thereuppon I sayd, as followeth:

Δ   — In nomine Jesu Christi, Quis tu es? — he answered
AN: Potestas omnis, in me sita est.   to Saul his
Δ   Qua?   hearing.
AN: Bona, et mala.
       Then appeared in the stone, these two letters M.G.
    I then asked him some questions, de Thesauro abscondito:
                                        he answered.
AN: Ne perturberis: Nam hæc sunt Nugæ.
                          And Jubtball appeared many child more skulls
                          on his left hand.
       she sayd to me,
AN: Ubi est potestas tua?
Δ   Cur quæris de potestate aliqua mea?
AN: Cur significas, non mihi placet.
Δ   I thereuppon, set by him, the Stone in the frame:
       and sayd
Δ   An bonus aliquis Angelus, assignatus est suæ speculæ?
AN: Etiam.
Δ   Quis?
AN: — [Hebrew] — he answered, by the shew of these letters ... Sto.
Δ   Bonus ne ille Angelus, de q in scripturis fit mentio?
AN: Maxime.
Δ   Fieri ne potest, quod ego eundem dicam, et cu illo agam?
AN: Ita.   and thereuppon appeared this character ___ Æ
Δ   Quid per hoc, significare velis?
AN: Alterius Angeli character est.
Δ   Cur hic, et nunc ostendis?
AN: Causam ob magnam — Make an ende.: It shalbe declared, but not by me
Δ   — By whome then
                                                     AN. — By him

crowded into the margins and inserted interlineally into the body of the diary entry.

Dee later described this conversation as one of their final conversations, but provided no information to suggest when the relationship had begun. Christopher Whitby argues that Saul began to scry for Dee in 1579 when a new prayer was put into use for contacting the angels.[30] Saul's known contact with Dee went back as far as 8 October 1581, when he was mentioned in Dee's private diary as a member of the household. Saul's place in Dee's household was clearly connected to inquiries into the occult properties of nature, and he might have gained his entrée specifically because of his reputation for seeing visions. On 9 October 1581, for instance, Dee recorded how "Barnabas Saul, lying in the . . . hall was strangely trubled by a spirituall creature abowt mydnight."[31] These visitations soon ended amid legal tangles and problems. Saul, like many of Dee's scryers, had trouble with the authorities and was taken to Westminster to face criminal charges on 12 February 1582.[32] A few weeks later the charges were dismissed and he returned to Mortlake to sever his scrying partnership with Dee. At that time Dee wrote that Barnabas Saul "confessed that he neyther h[e]ard or saw any spirituall creature any more."[33] Despite the cessation of spiritual activity, the break between Dee and Saul was never complete; intermittent entries in the personal diaries after March 1582 refer to Saul.[34]

In April 1586, while enmeshed in his troubles in Prague, Dee was coming to the end of his association with his second scryer, Edward Kelly. The two had been partners since March 1582, and angel conversations with Kelly dominate the existing diaries. Their first conversation took place on 10 March 1582, and their final extant conversation was dated 23 May 1587. Whether Kelly continued to scry between 23 May 1587 and Dee's departure from central Europe in the spring of 1589 is

30. Whitby, *AWS* I:50.   31. Dee, *PD*, p. 13.   32. Ibid., p. 14.
33. Ibid.
34. Saul's name was noted next to the date 19 February 1583, for example. This remark was omitted from Halliwell's edition of the diary. See Bodleian Library Ashmole MS 487, entry for February 1583.

Figure 1 (*opposite*). The first page of John Dee's surviving angel diaries, 22 December 1581. On the right side of the page, Dee drew a sketch of his stone, supported by a frame. On the left, the delta, or triangle, designates Dee; the angel's name, Anael, is marked "An." The scryer for this conversation was Barnabas Saul. British Library, Sloane MS 3188, f. 8ʳ. Reproduced by permission of the British Library.

not clear. During that time Kelly was under increasing pressure from
Rudolf II to perform lengthy alchemical experiments, and the time he
could have devoted to the angel conversations would have been limited.[35]

The facts surrounding Kelly's life are obscure. Various biographies
assert that he studied at Oxford under the name "Edward Talbot,"
trained as an apothecary, had his ears cropped at Lancaster for forgery,
and was a secretary to the bibliophile Thomas Allen.[36] Kelly, or "Edward
Talbot" as he was then known to Dee, was first introduced into the
household on 8 March 1582, only two days after Barnabas Saul removed himself from Dee's service.[37] Kelly was introduced to Dee by
"Mr. Clerkson," who had been to the house a week before bearing a
copy of a work by the medieval magus Albertus Magnus. Dee, Clerkson,
and Kelly dined together, and during the meal Kelly questioned Saul's
reliability as a scryer,[38] informing Dee that a "spiritual creature" told
him how Saul publically "censured" both Dee and Clerkson.[39] From
their very first meeting, Kelly presented himself as someone in communication with the occult levels of the cosmos and in possession of impor-

35. Dee remained familiar with Kelly's alchemical work, and it is impossible
to determine the extent to which he was a guiding force in the experiments.
Several references to alchemical results obtained by Kelly appear in Dee's
personal diary during this time: 15/25 January 1586; 25/5 September/
October 1586, 12/22 October 1586, 23/2 October/November 1586, 17/27
December 1586, 25/4 December/January 1586/1587, 2/12 January 1587,
19/29 January 1587. These entries have all been omitted from the Halliwell
edition. See Bodleian Library Ashmole MS 488, entries for January 1586
and September 1586–January 1587.

36. A summary of the biographical details of Kelly's life appears in Whitby,
AWS I: 43–51. Thomas Allen, one of the seventeenth century's great collectors, came into possession of at least twelve Dee manuscripts and may have
employed Edward Kelly before he entered Dee's service. Count Albert
Laski invited Allen to Poland in 1583, but, unlike Dee, Allen chose to
remain in England, where he continued to share intellectual interests with
Henry Percy, the earl of Northumberland. See Andrew G. Watson, "Thomas Allen of Oxford and His Manuscripts," in *Medieval Scribes, Manuscripts and Libraries. Essays Presented to N. R. Ker* (London: Scolar Press,
1978), pp. 279–316; Roberts and Watson, pp. 58 and 65.

37. Contrary to earlier beliefs that "Edward Talbot" and "Edward Kelly" were
different people, modern historians have adopted the position that one of
the two names was an alias. See French, *John Dee*, p. 113, n. 2, and Clulee,
p. 197. The earliest remarks concerning "Talbot" are in Dee, *PD*, pp. 14–
15.

38. Dee, *PD*, pp. 14–15.   39. Ibid., p. 15.

tant information about the natural world and Dee's often precarious place within it. Two days later, Kelly asked Dee for an opportunity to demonstrate his skill at "spirituall practise."[40]

Kelly acted as Dee's scryer at regular intervals until 4 May 1582, when an abrupt gap appears in the angel diaries and only a terse comment in Dee's personal diary helps to explain the disruption: "Mr. Talbot went."[41] Entries in the angel diaries do not resume until 15 November 1582. At some point during this period Kelly stopped being known under his alias, "Edward Talbot." Though the details of how Kelly's identity was uncovered are not known, one barely legible entry in Dee's personal diary from 6 May 1582 indicates that Jane Dee was in a "mervaylous rage" about Mr. Clerkson's reports of "cosening" or deceit – but the name of the individual concerned cannot be discerned. It is likely that it was Kelly, not only because of his relationship with Clerkson, but also because other barely legible notes in the personal diary during that period refer to him, including one where only "Talbot falsly" is legible.[42] Dee and Kelly reconciled on 13 July 1582 when the scryer visited Mortlake, still traveling under the name "Edward Talbot." Dee reported that the two men shared "some wordes of unkendnes" but "parted frendely."[43] The last reference to "Edward Talbot" appears in Dee's personal diary on 10 November 1582, but it was ruled out.[44] "Edward Kelly" emerged on 15 November 1582, when Dee began a new volume in his angel diaries entitled the "Quartus Liber Mysterioru[m]." On the title page Dee noted that the conversations took place after his reconciliation with Kelly.[45]

The long history of Dee's relationship with Kelly was no less troubled than his scryer's introduction into the household. Considering the disruptive effect Kelly had on Dee, his family, and the Mortlake community, the length and extent of their partnership is puzzling. Kelly stormed off in a rage at apparent insults, and Jane Dee was openly hostile to his presence.[46] Nonetheless, Dee was generous with personal and financial support for his scryer. The relationship between Dee and Kelly is even more difficult to fathom when we consider the agreement they made in 1587, at angelic behest, to share all things between them. This agreement specifically included the sharing of wives, which placed an unbearable strain on the already difficult relationship between the two men. Scholars

40. Dee, *AWS* II:16–19.   41. Dee, *PD*, p. 15.
42. Bodleian Library MS Ashmole 487, entries for May 1582.
43. Dee, *PD*, p. 16.
44. Bodleian Library Ashmole MS 487, entries for November 1582.
45. Dee, *AWS* II:138, where Dee noted "Post reconciliatione[m] Kellianam."
46. Harkness, "Managing an Experimental Household," pp. 254–255.

have speculated whether or not such sharing actually took place, but Dee wrote a brief, pointed entry in his personal diary – "Pactu[m] factu[m]" – suggesting that the agreement to share wives, bizarre as it was, was indeed physically consummated.[47] Within days of the pact the angel diary entries ceased, and the Dee–Kelly partnership was formally dissolved in December 1588. There was no bitterness on Dee's part, only disappointment. On 11 March 1589, Dee left his household in Trebona and traveled to Nuremberg while Edward Kelly remained in Prague.

Kelly's fortunes took steep upward and downward turns due to his alchemical expertise, which won him a permanent, though troubled, place at Rudolf II's court.[48] Alchemical ideas attributed to Kelly appeared in Bohemian alchemical manuscripts of the period, and specific experiments and processes bore his name as far away as Hungary.[49] Kelly's alchemical notoriety even led to the publication of *The Stone of the Philosophers* in 1590. This work indicated familiarity with the classic texts of early modern alchemy – the *Turba philosophorum* and works associated with Hermes Trismegistus, Arnold of Villanova, and Saint Dunstan – as well as the philosophical writings of Aristotle and Plato, and it is possible that the work represents the alchemical ideas of both Dee and Kelly. Dee certainly believed that *The Stone of the Philosophers* was not entirely Kelly's intellectual property, for he informed the archbishop of Canterbury on 13 July 1590 that he had certain rights "to the treatise of Sir Edward Kelley his Alchimy."[50]

Kelly's difficult personality ultimately proved his undoing: he was imprisoned in May 1591 in a castle near Prague for his reluctance to disclose alchemical secrets to Rudolf II.[51] After his arrest he continued to exercise an influence on Dee, whose sleep was troubled by dreams of his former scryer.[52] When Kelly's confinement was eased in 1593, Dee recorded the event in his personal diary: "Sir Edward Keley set at liberty by the Emperor."[53] Correspondence between the two men resumed, and Kelly reported that the emperor would once again welcome Dee's presence in Prague.[54] Kelly also had associations with the highest levels of English government during this period, for Dee received a "letter of

---

47. Both of these entries were omitted from Halliwell's edition. See Bodleian Library Ashmole MS 488, entries for May 1587.
48. Evans, *Rudolf II and His World*, p. 226–228.   49. Ibid. p. 226.
50. Dee, *PD*, p. 35.   51. Evans, *Rudolf II and His World*, p. 227.
52. Dee, *PD*, p. 44.
53. The news took two months to reach Dee. Dee, *PD*, pp. 46–47.
54. The letter arrived on 12 August 1595. Dee, *PD*, p. 53.

E.K." on 18 May 1594 that had passed through the hands of Queen Elizabeth.[55]

Kelly's death, like his life, is obscured by legends. It is possible that he was imprisoned again in 1595; he was not killed in an attempt to escape by jumping out of a tower window, though this story is often repeated. Although Dee received "newes that Sir Edward Kelley was slayne" on 25 November 1595, recent evidence suggests that reports of Kelly's death were greatly exaggerated.[56] According to R. J. W. Evans, he "was definitely alive in 1597 at the castle of Most," and there were reports of his activities as late as 1598.[57] A firm date for his death has yet to be established.

During the breakdown of his relationship with Kelly, Dee tried to use his son, Arthur (b. 1580), as a scryer. Arthur was first drawn into the angel conversations on 15 April 1587, after Kelly's repeated assertions that he no longer wished to scry for Dee.[58] Arthur's youth and virginity – two attributes traditionally prized in scryers – did not yield satisfactory results, however.[59] Despite an initial success when Arthur perceived two crowned men in the stone, Dee was frustrated by the lack of information his son was able to receive from the angels through the showstone. The rich sights and sounds Dee was accustomed to from Kelly's scrying were replaced by straightforward images – "squares, lines, pricks" and a few letters.[60] Three days after they began, Arthur and John Dee admitted defeat. No further suggestion was ever made regarding Arthur's scrying, nor is Arthur known to have commented on it in his later life.

Dee's final scrying relationship involved a long-standing acquaintance, Bartholomew Hickman (b. 1544).[61] Hickman is one of the angel conversation's most enigmatic figures, for if, as the evidence suggests, he scryed for Dee between 1591 and 1607, then it was he, not Kelly, who was Dee's most significant assistant. So few of the conversations involving Hickman survive, however, that it is impossible to get a full sense of the angels' revelations to Dee during the final years of his life. Their first

---

55. Halliwell interprets the entry as "letter of G. K." but the note actually reads "letter of E. K." See Dee, *PD*, p. 49; and Bodleian Library Ashmole MS 487, entry for May 1594.
56. Dee, *PD*, p. 54. 57. Evans, *Rudolf II and His World*, p. 227.
58. *TFR*, pp. *3–*8.
59. Thomas, *Religion and the Decline of Magic*, p. 230.
60. *TFR*, p. *7.
61. For a complete account of Hickman's life and his involvement with Dee, see Harkness, "The Scientific Reformation," pp. 342–346.

extant conversation, dated 20 March 1607, comes several decades after Dee mentioned Hickman in his private diary. Hickman first visited Dee on 22 June 1579 with his uncle and a friend, again on 12 July 1581, and then is not mentioned again until 1591, a few years after Dee returned from the Continent. Dee's personal diary contains regular references to Hickman up until 1598.[62]

Dee's relationship with Hickman was governed by Dee's belief that the summer of 1591 marked the beginning of a period of waiting and conversing with angels that was to last nine years and several months, and would culminate in a marvelous revelation or reward.[63] When the nine years and three months milestone was reached, Dee lamented that the period of waiting had been "co[n]futed as vayne."[64] On 29 September 1600 Dee manifested his disappointment by burning the angel diaries that he had kept during those nine long years. Despite this sobering incident, Dee and Hickman remained in close contact until the end of Dee's life, as is evident from the few angel conversations which Dee recorded in his aging and almost illegible hand and which still survive (Figure 2).

Although we may decide that Dee's relationships with all of his scryers – and especially Kelly – were based on chicanery and fraud, he nonetheless needed their assistance if his angel conversations were to succeed. Because scryers were an accepted, if sometimes frowned upon, part of the Elizabethan cultural landscape, there was no reason for Dee not to accept their services once he had decided to communicate with angels. That the angelic revelations changed little from one scryer to another confirms that it was *Dee* – the central participant – who lent coherence to the enterprise.[65] Without his influence and interests, which will be outlined in the following two chapters, the angel conversations would have collapsed from a lack of focus. Dee's natural philosophical interests

62. These references appear on 31 July 1591, 7 September 1592, 3 April 1593, 4 June 1593, 20 September 1593, 16–23 March 1594, 24–26 June 1594, 21–26 March 1595, 24 June 1595, and 26 October 1595. Not all of the entries regarding Hickman appear in Halliwell's edition of Dee's diary. Dee, *PD*, pp. 44, 46, 48, 50, 52, 55, 57; and Bodleian Library Ashmole MS 487, entry for July 1591.

63. See Bodleian Library, Ashmole MS 488, entries for September 1600.

64. Bodleian Library, Ashmole MS 488, entry for 25–29 September 1600.

65. See Harkness, "Shows in the Showstone: A Theater of Alchemy and Apocalypse in the Angel Conversations of John Dee," *Renaissance Quarterly* 49 (1996): 707–737, passim.

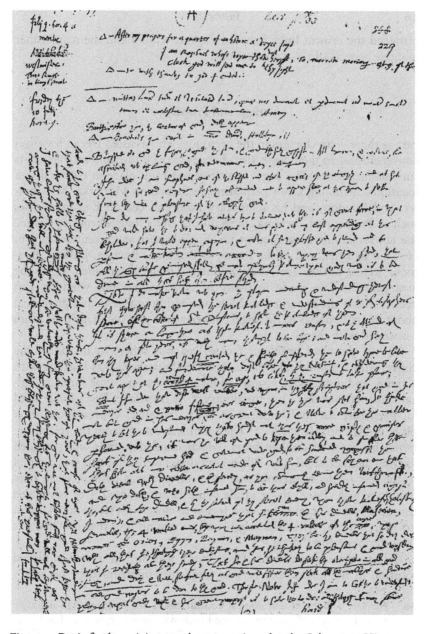

Figure 2. Dee's final surviving angel conversation, dated 2 July 1607. His scryer at that time was Bartholomew Hickman, but Dee's conventions for recording the conversation in his diary remain largely the same as in Figure 1. Note, however, that age did not improve Dee's handwriting. British Library, Cotton Appendix MS 46, f. 299ʳ. Reproduced by permission of the British Library.

provided the critical links between various scryers and the many other mutable aspects of the conversations.

Settings and Properties

Just as there were changes in the personnel who assisted Dee in his angel conversations, so there were also changes in the household arrangements surrounding them. The angel conversations took place in over a dozen locations, including Dee's permanent residences in Mortlake, Manchester, and London, and his temporary homes in central European cities such as Prague and Trebona. Even during Dee's long journey to Prague the conversations continued in the English port at Queenborough, in Holland at Briel, and in the towns of Dokkum, Bremen, Lübeck, Rostock, Stetin, Lasko, and Cracow. Little can be determined about the arrangements Dee made in many of these towns, but what we can deduce about the situations in Mortlake and Prague tells us a great deal about the parameters of his angel conversations and the many difficulties associated with them.

One of the chief problems facing Dee was where in his house the conversations should take place. The Dee household was bustling with activity, and unheralded visitors could stop by at any time for astrological consultations or to use the library.[66] At Mortlake, Dee's home until his departure for the continent in 1583, the angel conversations took place in his "private study."[67] The room was set off from the rest of the house by a set of double doors, which muffled sounds and provided varying degrees of privacy. When both doors were closed, it was a signal to the rest of the household that strict privacy was desired; when only one door was closed, those inside could be interrupted. Dee sabotaged this neat system because he often forgot to close both study doors, as on 20 March 1582, when the angels had to remind him that he "had left the uttermore dore" open.[68]

Despite these precautions, Dee's study was still in a well-trafficked location in the house, and he was frequently interrupted by his wife's

---

66. For Dee's diary of astrological consultations, see Bodleian MS Ashmole 337, ff. 20–57. His clients included such well-known figures as Lord Herbert and Lady Blount, ff. 49v and 50v. For a discussion of the role Dee's library played in the intellectual life of England and the visitors there, see Sherman, *John Dee*, especially pp. 38–41.
67. See Dee, *AWS* II:17, 218, 301, and 326.   68. Ibid., p. 63.

calls to dinner and by unexpected visitors. On one occasion, a messenger from the court burst into the private study in the middle of an angel conversation. Dee, "fearing his rash opinion afterward of such things, as he could not perceive perfectly what my Companion [Edward Kelly] and I were doing," anxiously asked the angels for advice on how to minimize any possible damage to his reputation.[69] Privacy remained of grave concern to Dee and his associates throughout the angel conversations, and Dee asked the angel Michael on 4 May 1582 if he and Kelly might use the "furdermost little chamber ... yf the bed be taken downe" instead of the private study.[70] This arrangement would have helped to diminish interruptions as well as safeguard Dee's reputation from charges of conjuring, since the chamber in question was the last room in a series of linked rooms and thus more completely removed from the household.

The quiet and relative isolation Dee craved was an important feature of many early modern natural philosophical programs. Isolation was especially important in the angel conversations because of the traditional belief that solitude was necessary for communication with God – or, at the very least, that solitude increased the likelihood that such communication would take place. Yet Dee, like most early modern natural philosophers, was eager to acquire a battery of noble patrons and financial sponsors. Such individuals could not be found in Dee's study or his "furdermost little chamber"; rather, potential patrons gravitated from their country estates to the court, and Dee's home at Mortlake provided an ideal rest stop between the great estates in the south and west of England and the royal palaces on the river Thames such as Greenwich and Whitehall. Incidents like those described – from the interruption of servants to the possible misinterpretation of events by those not as erudite as himself – demonstrate some of the difficulties associated with Dee's attempts to reach a compromise between the contemplative life that his angel conversations demanded and the active, political life of a court-sponsored natural philosopher upon which his financial well-being depended.[71]

Dee continued to grapple with the tension between public and private throughout his life and employed a number of strategies to help accommodate the special needs of a contemplative, financially solvent Christian

---

69. *TFR*, p. 23.   70. Dee, *AWS* II:137.
71. For a detailed interpretation of these conflicting ideals in a slightly later period, see Stephen Shapin, "The Mind Is Its Own Place": Science and Solitude in Seventeenth-Century England," *Science in Context* 4 (1990): 191–218.

natural philosopher.[72] One strategy involved a small oratory – a place for private devotions and contemplation – next to the study.[73] Dee retreated to his oratory before each angel conversation to pray for divine favor and guidance. This space was probably similar to Heinrich Khunrath's oratory as depicted in the *Amphitheatrum sapientiae aeternae* (1602). In Khunrath's study, which contained everything from alchemical vessels to musical instruments and books, the prominent placement of the oratory in the foreground of the illustration implies that prayer was integral to the work of alchemy. The angel conversations demanded a similar space for contemplation and meditation.[74]

At no time were the problems associated with blending the active and contemplative lifestyles more apparent than when Dee was traveling in Europe from 1583 to 1588. During this period the physical arrangements surrounding the angel conversations were changed frequently, and finding suitable accommodations was especially difficult. Any potential residence had to have isolated rooms for the conversations, rooms suitable for alchemical experiments, and a location that was accessible to both the Court and suppliers of alchemical materials. When Dee first arrived in Prague in August 1584 he was able to lease such a house from another natural philosopher, Tadéas Hájek. Though Hájek is best known today for his astronomical work, he was also interested in alchemy and was a notable figure at Rudolf's court. Among its other attractions, Hájek's house contained a study decorated with alchemical friezes. Dee believed it to have been "in times past . . . the Study of some Student . . . skilfull of the holy stone."[75] A name – "Simon Baccalaureus Pragensis" – was inscribed on the walls in silver and gold, and among the pictures Dee mentioned "very many *Hieroglyphical* Notes *Philosophical*, in Birds, Fishes, Flowers, Fruits, Leaves, and six Vessels, as for the Philosophers work." Two literary inscriptions within the friezes described the alchemical art.[76]

72. See Deborah E. Harkness, "Managing an Experimental Household: The Dees of Mortlake," *Isis* 88 (1997): 242–262.

73. The oratory was a separate room rather than a secluded place within the study. One of the participants in the angel conversations was able to exit from the study by passing through Dee's oratory. *TFR*, p. 23.

74. See the illustration of his "Laboratium oratorium" in Heinrich Khunrath. *Amphithearum sapientiae aeternae* (Hanau, 1609).

75. *TFR*, p. 212.

76. Ibid. Verses over the door read: "Immortale Decus par gloriaque illi debentur/Cujus ab ingenio est discolor his paries." A longer passage was written on the uppermost portion of the south wall of the study:

Hájek's house proved too small for the growing number of people and activities associated with Dee's projects, and in January 1585 Dee moved to a house on Salt Street near the Old Market in Prague. This house was better designed for the complexities of Dee's natural philosophical practices as it had two studies where the angel conversations could take place: a "mystical study," and a "secret study."[77] Once again the household's activities proved distracting, and by April 1586 Dee had withdrawn to an even more isolated location in the house's tower when he wished to communicate with angels. The tower contained an "oratory . . . at the top . . . a small heated room, truly elegant and commodious."[78] The oratory or private space for prayer thus remained a vital component of the changing environments where the angel conversations took place even after Dee left Mortlake.

Despite the mutability of the angel conversations' larger environment, their integrity depended upon a more easily defined, portable set of properties: the showstone, the Holy Table, and the Holy Books. For Dee, the focus of his angel conversations was not his library, alchemical laboratory, study, or oratory, but his showstone. He used at least three showstones in the angel conversations: a "great Chrystaline Globe," a "stone in the frame," and a stone Dee believed had been brought by the angels and left in his oratory window at Mortlake. The last was the most powerful, and only Dee was allowed to touch it.[79] He set this

---

Candida si rubeo mulier nupta sit marito: Mox complectuntur, Complexa concipiuntur. Per se solvuntur, per se quandoque perficiuntur: Ut duo quae fuerant, unum in corpore fiant: Sunt duae res primo, Sol cum Luna, Confice, videtis, sit ab hiis lapis quoque Rebus.

Lunae potentatu, peregit Sol Rebis actu: Sol adit Lunam per medium, rem facit unam. Sol tendit velum, transit per ecliptica Caelus: Currit ubi Luna recurrit hunc denuo sublime. Ut sibi lux detur, in sole quae retinetur. Nec abiit vere, sed vult ipsi commanere: illu irans certe defunctum copus aperte: Si Rebus scires, quid esset tu reperires. Haec ars est cara, brevis, levis atque rara. Art nostra est Ludus puero, labor mulierum; scitote omnes filii artis huius, quod nemo potest colligere fructus nostri Elixiris, nisi per introitum nostri lapidis Elementati, et si aliam viam quaerit, viam nunquam intrabit nec attinget. Rubigo est Opus, quod fit ex solo auro, dum intraverit in suam humiditatem.

77. The "mystical study" is mentioned in *TFR*, p. 367; the "secret study" is mentioned months later, in *TFR*, p. 409. It is possible, but unlikely, that the two studies were actually the same space.
78. Bodleian Library MS Ashmole 1790, ff. 1–10. Dee in Josten, p. 240.
79. Whitby, *AWS* I: 137–141; Dee, *AWS* II:217–218.

angelically delivered stone in gold and intended to wear it around his neck.[80]

A fourth "stone" associated with Dee's angel conversations is the black obsidian mirror currently owned by the British Museum. Based on the evidence from the angel diaries, this mirror was not used in the angel conversations. Frequently, Dee made sketches of the stones in the margins of his manuscripts, none of which, however, portrays the black obsidian mirror (see Figure 1). What is more, there is no direct evidence that the mirror belonged to Dee. The legend associating the black mirror with the angel conversations was based on the 1748 catalogue of Horace Walpole's collection of curiosities. In the catalogue, the mirror was described as "a speculum of kennel-coal, in a leathern [sic] case. It is curious for having been used to deceive the mob by Dr. Dee, the conjurer, in the reign of Queen Elizabeth."[81] Here, Dee's angel conversations are not even mentioned. It was in the nineteenth century that Francis Barrett reported, in his compilation of magical materials entitled *The Magus* (1801), that "six or seven individuals in *London* . . . assert they have the stone [used in the conversations] in their possession."[82] Somehow the two legends intertwined and the stories reporting Dee's use of the obsidian mirror in the conversations have been repeated for several centuries on the basis of only "very circumstantial" evidence.[83]

Dee's use of the showstone represented the most traditionally occult aspect of his angel conversations.[84] Dee noted some of its connotations in a presentation copy of selected angel conversations, *De heptarchia*

---

80. *TFR*, p. *40. A current provenance for the stone is not available, but it is probable that such a precious object was buried with him.

81. Quoted in Hugh Tait, "The Devil's Looking Glass," in *Horace Walpole: Writer, Politician, and Connoisseur*, ed. Warren H. Smith (New Haven and London: Yale University Press, 1967), pp. 195–212; see p. 200.

82. Francis Barrett, *The Celestial Intelligencer* (London, 1801), p. 196.

83. Goldberg, for instance, repeats the long-standing belief that Dee used the obsidian mirror in the conversations. See Benjamin Goldberg, *The Mirror and Man* (Charlottesville: University of Virginia Press, 1985), pp. 17–18. Whitby is more cautious and posits that the mirror might have been owned by Dee, and perhaps even used in the conversations, but that there is no evidence to support this contention. See Whitby, *AWS* I: 138–141.

84. See Goldberg, *The Mirror and Man*, especially pp. 3–162. Relevant medieval and early modern ideas concerning stones and their relationship to magic and astrology are discussed in Carla de Bellis, "Astri, Gemme e Arti Medico-magiche nello "Speculum Lapidum" Di Camillo Leonardi," in *Il Mago, Il Cosmo, Il Teatro degli Astri: Saggi sulla Letteratura Esoterica del Rinascimento*, ed. Gianfranco Formichetti, with an introduction by Fabio Troncarelli (Rome: Bulzoni Editore, 1985), pp. 67–114.

mystica. In this volume, he noted the work of Ephodius and Epiphanius on precious stones, as well as an unspecified "book received at Trebona."[85] He also drew attention to Epiphanius's remarks about "Adamanta," a stone "in which diverse signs are given for responding to God," and to the Jewish belief that the original laws given to Moses on Mount Sinai "were expressed in sapphires."[86] The Jewish mystical text *Zôhar* supported this belief, recounting the story that the original tablets of the law were constructed from a divine sapphire that served as the foundation of the cosmos after being cast down from heaven by God. Moses destroyed the original sapphire tablets and copied them into rough stone before sharing them with his people. Thus, whereas the original sapphire tablets conveyed the pure, eternal truths of God, Moses' tablets were only imperfect imitations of that truth.[87]

In Dee's conversations the showstone clearly served as a locus for divinity, a sacred space capable of representing the world in microcosm. Dee noted that "the whole world in manner did seme to appeare, heven, and erth. etc." within the stone.[88] At times, a more specific microcosm – the study where the conversation took place – appeared in the crystal showstone, along with Dee and his scryer. This microcosm within a microcosm would quickly evaporate, yielding to a representation of the "whole world."[89] The angels made explicit references to the safe, sacred nature of the showstone, for it served as a protective device to keep potentially evil spirits at bay. The angels assured Dee that, within the showstone, "*No unclean* thing shall prevayle."[90] Instead, the stone could contain only true, holy angels.[91]

85. A printed edition of this work is available. See John Dee, *The Heptarchia mystica*, edited and annotated by Robert Turner (Wellingborough, Northants.: Aquarian Press, 1986), p. 64. There is no reference to a work by "Ephodius" in the Roberts and Watson catalogue; Dee did own Epiphanius's work on the breastplate of Aaron, contained in a 1565 collection of works on stones and gems including Johannes Kentmannus's *De fossilibus* and other tracts by authors such as Conrad Gesner. See Roberts and Watson, #765; this volume has not been recovered, and was part of Dee's continental traveling library.

86. Dee, *Heptarchia mystica*, p. 64.

87. A. E. Abbot, *Encyclopaedia of the Occult Sciences* (London: Emerson Press, 1960).

88. Dee, *AWS* II:48. Caret marks indicate where Dee has inserted text above the line.

89. *TFR*, p. 115.   90. Dee, *AWS* II:247 (Dee's emphasis).

91. Simon Forman, one of Dee's contemporaries and a natural philosopher with an interest in the occult sciences, noted similar ideas about the sacrosanct nature of stones used to contact angels in his book of angelic experi-

Dee's ability to communicate with angels through his sacred show-stone might have been linked by contemporaries to his expertise in alchemy because of analogies drawn between the showstone and the closed vessel of an alchemical experiment. Dee believed that the show-stone was linked to his alchemical success and noted in his copy of Marcus Manilius's *Astronomica* (1533) the associations between the extraordinary new star of 1572, his alchemy, and the angel conversations:

> I did coniecture the blasing star in Cassiopeia appering a° 1572 to signify the fynding of some great Thresor or the Philosophers stone. . . . This I told to Mr. Ed. Dier. at the same tyme. How truly it fell out in a° 1582 Martij 10 it may appere in tyme to come ad stuporem Mundi.[92]

The date – 10 March 1582 – marked Dee's first angel conversation with Edward Kelly. Yet there is no mention of alchemy in that conversation. The connections Dee made between these seemingly disparate events suggest that an existing, perhaps oral, tradition supported the connection. After Dee's death Elias Ashmole (1617–1692), a seventeenth-century scholar interested in both alchemy and Dee's angel conversations, wrote in his *Theatrum Chemicum Britannicum* (1652) that the elusive philosopher's stone, when subjected to even higher and more mystical processes, yielded "vegetable," "magical," and "angelic" stones, each characterized by its own specific properties and powers.[93] The most exalted, the angelic stone, was a substance so subtle it could not be seen, felt, or weighed, but only tasted. Ashmole described the stone in some detail:

> It hath a *Divine Power, Celestiall,* and *Invisible,* above the rest; and endowes the possessor with *Divine Gifts.* It affords the *Apparition* of *Angells,* and gives a power of conversing with them by *Dreames* and *Revelations:* nor do any *Evil Spirits* approach the *Place* where it *lodgeth.* Because it is a *Quintessence wherein there is no corrupt-*

ments. Forman recorded a warning not to use the same crystal stone to converse with both holy angels and simple (possibly evil) spirits, writing that "ye must never call any spyritte into that Stone you call Aungells into." See Forman, "Magical papers," British Library, Add. MS 36674, ff. 47–56ᵛ, f. 56ᵛ.

92. Marcus Manilius's *Astronomica* (Basel, 1533), book V. Dee's copy is now University College, London Ogden A.9. For more information, see Roberts and Watson, #251.

93. Elias Ashmole, "Prologomena," *Theatrum Chemicum Britannicum* (London, 1652), sig. A4ᵛ.

ible Thing: and where the Elements are not corrupt, no Devil can
stay or abide.[94]

Ashmole believed that Hermes had knowledge of the angelic stone which he refused to share with anyone, and that the only other people to possess it were Moses and Solomon.[95] Thus, the connections Dee made between alchemy and angels were not unique but part of an extant tradition that helped to join the two main intellectual projects in Dee's later life: his alchemy and his angel conversations.

A stone as sacred and significant as Dee's required a proper setting, so he used a special table and wax seals to further delineate the sacred space surrounding his showstone. Though the showstone was not dependent upon its environment to function properly, it was nonetheless a holy object to be revered and set apart from the mundane world. Because the stone was so crucial to Dee's angel conversations, the diaries record specific information about the furnishings, personal items, and books that complemented it. During each angel conversation Dee placed the showstone on a large wax seal, which was itself resting on a "Holy Table." Together, these items provided a larger sacred space with increased religious and occult connotations. Dee's use of a wax seal, for example, was related to medieval and early modern Solomonic magical practices involving the use of sigils and talismans, and his "Holy Table" served as an altar adorned with figures and inscriptions.[96] The designs for these objects did not come from standard occult references but from the angels themselves.

The "Holy Table" was, after the showstone, the most important property Dee used in the angel conversations. The angels gave initial designs for the table early in the conversations, on 10 March 1582.[97] As in the case

94. Ibid., sig. B^v.
95. Ibid., sig. B2^r. I was unable to find any mention of this stone in the texts of the Hermetic corpus; see the translation by Brian P. Copenhaver, *Hermetica: The Greek "Corpus Hermetica" and the Latin "Asclepius"* (Cambridge: Cambridge University Press, 1992), passim.
96. E. M. Butler, *Ritual Magic* (Cambridge: Cambridge University Press, 1949), contains a detailed discussion of the Solomonic magical tradition on pp. 47–99. Richard Kieckhefer's *Magic in the Middle Ages* (New York: Cambridge University Press, 1990) discusses the medieval use of sigils and talismans on pp. 75–80. For the possible influence of Solomonic magic on Dee, see Stephen Clucas, "John Dee's *Liber Mysteriorum* and the *ars notoria*: Renaissance Magic and Medieval Theurgy," in Stephen Clucas, ed., *John Dee: Interdisciplinary Studies in Renaissance Thought* (Dordrecht: Kluwer Academic Press, forthcoming).
97. Dee, *AWS* II:22–23; Whitby, *AWS* I:150–154. To make a composite description, Whitby used information from both the angel diaries and the

of most of the material objects Dee was ordered to make, a complete understanding of the table's construction and use is not available. It is not
known who made the table for Dee, and he struggled to clarify the angels'
instructions for some time. He was particularly concerned about how to
replicate divinely mandated objects in the imperfect sublunar world. Finally the angels soothed some of Dee's anxieties, telling him that the table
used in the angel conversations was only an earthly representation of a divine, celestial table, which the angel Uriel explained was "A Mysterie, not
yet to be known."[98] The perfect, holy table shown in the angels' revelations was only symbolized by his personal table, and he may have conceived of the other objects used in his angel conversations in a similar fashion – as earthly representations of eternal, divine objects.

Once the table was made, Dee faced another problem: how to trans-
port it to the court of Rudolph II. Because the table was the largest
object used in the conversations, hiding it from curious eyes was difficult.
Actually carrying it might have been still more challenging. Dee considered having a new table made in each city, but this potential solution
was discarded for reasons of time and secrecy.[99] Instead, the table was
especially designed for travel. When Elias Ashmole saw Dee's "Holy
Table" in the seventeenth century, it was actually a portable, flat tabletop enclosed in a protective wooden case with iron locks. When in use,
the table rested on a wooden frame that raised it a little over two and
half feet from the ground.[100] Apparently Dee had a new frame built in
each city because, after some time traveling, he became concerned that
to have a new base made for the table in each location still jeopardized
his entourage's safety. Once they reached their patron's estates in Lasko,
Dee asked the angels if he could take the base being made there to
Cracow, "both to save time, and to have our doings the more secret."[101]

Although the angels did not satisfy Dee's requests for more detailed
instructions regarding the table's base, they did give exact specifications
for the tabletop. The angels told Dee to make the tabletop from a four-
feet-square piece of sweet wood (though Ashmole claimed the table he
saw in Cotton's library was only about three-feet square).[102] The tabletop might have been painted by the painter and engraver Richard Lyne
(fl. 1570–1600), with whom Dee conferred around the time of the

testimony of Elias Ashmole, who saw the table in John Cotton's library in
the seventeenth century. I have drawn from his synthesis in these para-
graphs. There were actually two sets of instructions delivered to Dee concerning the table. The earlier ones, given on 10 March, were later labeled
"false," and new instructions followed on 26 April 1583.

98. Dee, *AWS* II:21.   99.   *TFR*, p. 69.   100.   Whitby, *AWS* I: 151.
101.   *TFR*, p. 69.   102.   Dee, *AWS* II:21.

table's construction.[103] The top was decorated with a large central square containing letters representing angelic names drawn from a mystical alphabet the angels had shared with Dee. A pentagram and seven smaller emblems known as the "seven Ensigns of Creation" surrounded the central device.[104] According to the angels' initial instructions, the "Ensigns of Creation" were to be inscribed in tin and placed on the table, but later Dee was permitted to paint them on the surface using blue, red, and gold pigment.[105] Characters from the angels' mystical alphabet provided a blue border for the table, with gilding at the corners for extra ornamentation. The completed table must have looked like an altar, with its central, elevated showstone and magnificently decorated surface.

The altarlike appearance of the table was accentuated by Dee's use of cloth draperies and wax seals. The angels instructed him to place red silk under the table legs and to position another piece of silk on top of the table, covering the wax seal and frame positioned under the showstone.[106] This arrangement elevated the showstone above the rest of the table, increasing the distance between the divine showstone and the more mundane materials used in the angel conversations. Like the showstone, the wax seal was a physical link between the divine and the human, the celestial and terrestrial, and the angel Uriel instructed Dee to look upon the seal with "great reverence and devotion."[107] The large wax seal was known as the "Sigillum Dei," whose proper name, the angels told Dee, was "Emeth" (Figure 3).[108] *Emeth*, the Hebrew word for "truth," was one of the seventy-two names of God in the cabalistic tradition.[109] Be-

---

103. Ibid., p. 352. In a discussion with the angel Il on the significance of proportion, Dee referred to "master Lyne" and asked whether he could "serve the turn well?" before he showed the angel a rough draft of the letters to be painted around the table's edge. Lyne had connections – as did Dee – to Cambridge University and Dutch intellectual circles, and was well known for his maps, genealogical tables, and portraits. See the *DNB*, vol. 12, p. 342, and Francis Meres, *Palladis Tamia* (London, 1598), p. 287ᵛ.

104. Christopher Whitby believes the pentagram and ensigns related both to the seven days of God's creation and to the seven stages of alchemical transformation. Whitby, *AWS* I:134–135.

105. Ibid., pp. 134–137; Dee, *AWS* II:383–384.

106. Dee, *AWS* II:22. The angels later altered these instructions, requesting that Dee procure a "changeable" or shot silk. See ibid., p. 379.

107. Ibid. p. 21.

108. Ibid., pp. 32–33; Whitby, *AWS* I:118–119. Dee made a marginal note to remind him to investigate both Reuchlin and Agrippa for this name, and specifically mentioned Agrippa's *De occulta philosophia* III, chap. 11.

109. Agrippa, *DOP*, p. 227.

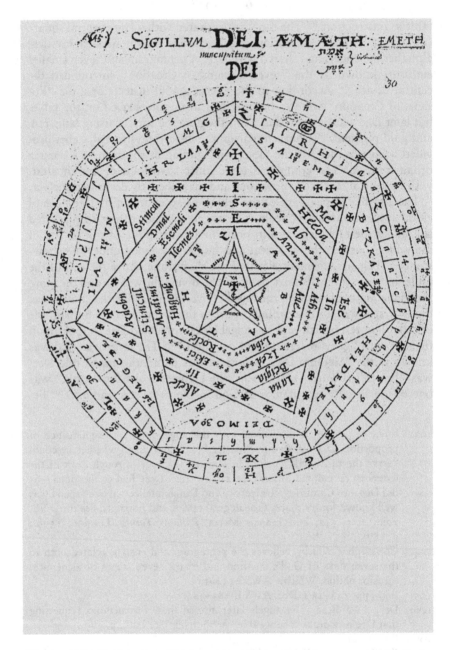

Figure 3. The Sigillum Dei, which was inscribed and used as a support for Dee's crystal showstone. British Library, Sloane MS 3188, f. 30ʳ. Reproduced by permission of the British Library.

tween 10 and 21 March 1582 the angels instructed Dee to make the seal from undyed beeswax and to inscribe the seal with a heptagon, letters from the angelic alphabet, names of the angels, and the secret names of God.[110] On the reverse, a cruciform symbol was inscribed with the word "AGLA." In the Middle Ages AGLA was one of the names of God used in exorcisms, healing magic, and divination.[111] The mediating and protective function of the larger wax seal was enhanced by four smaller versions of the "Sigillum Dei." Each was encased in a metal canister and placed under one of the legs of the Holy Table. Each angel also conveyed his personal seal or talisman to Dee.[112]

The showstone, the table, and the wax seals were the central physical features of the angel conversations. They demarcated Dee's "angelic laboratory" where divine communication could take place. Dee's own person was also important to the integrity of the conversations, and the angels described objects he was to wear, including a ring and a triangular "lamine," or breastplate, inscribed with symbols. In addition to the ring and lamine, the angels told him to make a special rod named "EL." But, according to the angel diaries, Dee's use of the ring, lamine, and rod was never central to the angel conversations in the way that the stone, table, or seals were.[113] Instead, they were objects that sanctified Dee, symbolized his role as a conduit for divine influences, and protected him from evil. The angels described no objects that the scryers were to possess or wear, and their absence underscores that it was Dee, rather than his assistants, who played the crucial role in the conversations.

Dee received detailed instructions for both the ring and the lamine. According to the angel Michael, the ring was to be made of precious metals and inscribed with the name "PELE."[114] Dee made a note to check for the name "PELE" in his most authoritative Christian cabalistic text, Johannes Reuchlin's *De verbo mirifico*.[115] PELE did not appear

110. See Whitby, *AWS* I:119–123 for a complete discussion of how to decode the names hidden within the seal.

111. Kieckhefer, *Magic in the Middle Ages*, pp. 73, 85, and 89.

112. Like the use of the word "AGLA," the tradition of angelic seals or sigils was long-standing in occult texts. An English manuscript dated 1562, for instance, included a seal containing the names of six angels: Armediel, Darionel, Ramaniel, Gabriel, Carbrael, and Ranael. Cambridge University Library, MS Li.1.12, f. 21ᵛ. Of these angels, only Gabriel appears in Davidson's *Dictionary of Angels*; the rest are idiosyncratic, like many of Dee's angels.

113. Whitby, *AWS* I:131. This assertion must be qualified, for, as in the case of the angelic stone set in gold, Dee may well have been buried with the objects.

114. Dee, *AWS* II:31–33.   115. Ibid., p. 32.

there, but could be found instead in Agrippa's *De occulta philosophia.*
Dee went on to insert PELE in the margins of Dionysius the Areopagite's
"Divine Names" next to a passage that mentioned the "wonderful 'name
which is above every name.' "[116] Dee's interest in the ring increased when
the angel Michael asserted that its design had not been "revealed since
the death of Salomon," the archetypal biblical figure of wisdom.[117] Michael went on to describe how Solomon's ring enabled him to perform
his "Miracles, and divine works and wonders."[118] Similarly, Dee needed
the ring to perform the tasks God would set him. *"Without this [ring],"*
Michael told Dee, *"thow shalt do nothing."*[119]

The "lamine," or breastplate, was a companion piece to the ring. First
described on 10 March 1582 as a thin, triangular, metal plate of gold
inscribed with sigils and worn around the neck, the heritage of Dee's
lamine, like the wax seals, can be found in the medieval talismanic
tradition associated with Solomon. In the case of the lamine, however,
the angels' instructions suggest a closer link to traditional amulets used
against illness, evil spirits, and bad fortune, since the angel Uriel told
Dee that it would protect him from danger.[120] The lamine's design was
more symbolically charged for Dee than the ring. First, the shape was
significant. Dee had been using the triangle as a personal hieroglyphic
for some time, because the Greek delta or letter D provided a reference
to his name. In addition, the triangular lamine was a symbolic reference
to the Holy Trinity of the Father, Son, and Holy Spirit, which enhanced
its divine connotations. Dee also related his angelic lamine to the Urim
and Thummim that ancient Hebrew priests and prophets used for divination in matters important to the community (Exodus 28:15–30; Levit-
icus 8:8; Numbers 27:21).[121]

116. See ibid. Whitby, *AWS* I:131; Dionysius the Areopagite, Roberts and
   Watson, #975, p. 146[r]: "Ac veluti ab omni eum divini nominis cognitione
   abduce[n]tem dixisse cur de nomine me interrogas quod *est mirabile.* An
   vero istud non est *mirabile* nomen: quod est super omne nomen, quod est
   sine nomen, quod omne exsuperat nomen . . ." (Dee's emphasis). Against
   this passage, Dee noted "Pele."
117. Dee, *AWS* II:31.   118. Ibid.   119. Ibid, p. 32; Dee's emphasis.
120. "Sigillum hoc in auro sculpendum, ad defensione[m] corporis, omni loco,
   tempore et occasione. et in pectus gestandum." Dee, *AWS* II:19. See also
   Whitby, *AWS* I:129–131, and Kieckhefer, *Magic in the Middle Ages,*
   pp. 77–80.
121. See also Dee's emphasis marks in his copy of Sebastian Muenster, *Messias
   Christianorum et Iudaeorum Hebraicè & Latinè* (Basel, 1539), Roberts
   and Watson, #1616. "Fuerunt pr[a]etera & aliae quinq[ue] res in te[m]plo
   primo, quae non fuerunt in secundo, nempe h[a]e arca cum propriciatoro

Other objects of crucial importance to Dee's angel conversations were his angelically delivered books, which can be divided into two broad categories. First, Dee recorded the day-to-day angelic revelations in notebooks he maintained chronologically, here referred to as "angel diaries." These contain information about events that took place in the household during the angel conversations, as well as the angels' revelations. The second class of books – the revealed books – were given to Dee in the course of the conversations, and they were always meant to be extracted from the chronologically recorded diaries and maintained as separate books. Dee's extant copies of the revealed books – the *48 Claves angelicae, Liber scientiae auxilii et victoriae terrestris, Tabula bonorum angelorum invocationes* [sic], *De heptarchia mystica*, and *Liber Logaeth* – were finely executed in his most polished handwriting.[122] Their quality of presentation suggests that they were intended for reference and possibly as gifts to patrons and associates, rather than being rough notes of the angel conversations.

As with most of the objects associated with Dee's angel conversations, his "holy books" represent a mixture of natural philosophical, occult, and religious traditions, a unification made possible because of the power associated with language in Judeo-Christian creation accounts in which God used speech to craft the cosmos. During Dee's lifetime, Protestant reformers extended the living power of God's speech to include books of scripture, which they argued should be revered as if they contained the spoken word of God. The sixteenth-century theologian John Calvin remarked that "all of Scripture is to be received as if God were speaking," and any records of divine speech, such as the angels' revelations to Dee, would have been considered authoritative.[123]

The Scriptures were not the only powerful texts, however. In the Judeo-Christian and Islamic traditions "holy books" containing divine and angelic revelations were equally powerful. The tradition of holy

& Cherub, ignis coelitus missus, *spiritus sanctus, Urim & thumim,* idest, *mysteria qu[a]eda[m] sacra in pectorali summi sacerdotis.*" For an early modern translation, consult Sebastian Muenster, *The Messias of the Christians and the Jewes,* trans. Paul Isaiah (London, 1655).

122. These volumes were found in a secret compartment of a storage chest belonging to Dee. They are now bound as British Library, Sloane MS 3191. A published edition of these texts appears as John Dee, *The Heptarchia mystica* edited and annotated by Robert Turner (Wellingborough, Northants.: Aquarian Press, 1986), passim.

123. Quoted in William A. Graham, *Beyond the Written Word: Oral Aspects of Scripture in the History of Religion* (Cambridge: Cambridge University Press, 1987), p. 143.

books in these religions can be traced back even further to Near Eastern and Mediterranean examples. Typically, ancient holy books took one of several forms: books of wisdom, books of destinies (containing the knowledge of human lives and history in the future), books of works (recording human deeds in anticipation of final judgment), and books of life (containing the names of the elect to be saved at the final judgment).[124] Dee viewed all holy texts – be they scriptural authorities or private revelations – as having inherent power, highlighting in his copy of Azalus's encyclopedia that magic existed even in the written accounts of the Christian patriarchs.[125] He considered his diaries a set of similar, extracanonical revelations that could be consulted, along with existing books of holy scripture contained in the Bible, for answers to his questions about nature and the world.

Within the occult tradition books possessed additional powers, from the *Clavicles* associated with Solomon to practical magical manuals known as *grimoires*. Studying these texts contributes little, however, to an explanation of Dee's detailed diary entries.[126] Some grimoires were simply catalogues of rituals and prayers, for example, while others contained technical descriptions of materials, diagrams of talismans and sigils, and detailed procedures. Still other early modern occult books consisted of long lists of beneficent and maleficent spirits, including the forms they took, their powers, and their attributes.[127] Even other contemporary records of conversations with angels, spirits, and demons are not analogous to the diaries that record Dee's angel conversations. Most comparable were the rituals employed by Dee's contemporaries "HG" and "Jo. Davis" on 24 February 1567, when a holy book was used to facilitate visions of angels in a stone who revealed additional sacred books.[128] Though occult practitioners had been calling spirits into the world for centuries to look for buried treasure and act on their wishes, few asked those spirits for information about the way the world worked, or how and when God would reorder it according to a divine plan. In Dee's angel conversations, by contrast, these are central concerns.

Dee's angelically revealed books were delivered in a sequence that

124. Ibid., pp. 50–51.
125. Pompilius Azalus, *De omnibus rebus natmalibus* (Venice, 1544), Roberts and Watson, #134/B188. p. 19ʳ. "Et de ipsor[um] formis in *vita* patru[m], & aliis scripturis sacris plures lengunt[ur], tanq[uam] *magica* co[n]sistent[ur]"; Dee's emphasis.
126. E. M. Butler, *Ritual Magic* (Cambridge: Cambridge University Press, 1949), pp. 47–88.
127. Kieckhefer, *Magic in the Middle Ages*, p. 170.
128. "Magical Papers," British Library, Add. MS 36674, ff. 59ʳ-62ᵛ.

went from grammatical and linguistic basics to complicated matters of religious doctrine. The first revealed book conveyed to Dee and Kelly was the *Liber Logaeth*, or "the Book of the Speech of God."[129] It conveyed the angelic language and emphasized the role that language would play in perfecting the world and returning it to its original state. The *Liber Logaeth* consisted of forty-nine "calls" or prayers and ninety-five tables containing small square spaces filled with letters (Figure 4). The angels delivered additional prayers as a preamble to the *Liber Logaeth*, but they are not included in the finished text.[130] The angels instructed Dee to complete the book forty days after 29 March 1583, a period of time that had both biblical and alchemical significance.[131] The angels promised that the book and the Holy Table would, when used at God's appointed time, bring about the apocalyptic redefinition of the natural world.

The second book the angels revealed, *De heptarchia mystica*, was a reference summary of information conveyed to Dee before he departed for the Continent in 1583.[132] It contained prayers to specific angels that were to be repeated at different times of the day and on different days of the week, as well as the seals for forty-nine elemental angels with powers over aspects of the natural world.[133] Because the angels delivered this information at various points throughout the angel conversations, Dee would have found it difficult to find specific instructions. By drawing the materials together in a separate volume, however, Dee provided himself with an encylopedic survey of key information revealed in the conversations. Although Whitby contends that the book provided "Dee with ready access to all that he needed for the practical summoning of these spirits," there is no evidence that Dee ever used the prayers in his daily practices.[134] Instead, he used traditional, orthodox prayers at the beginning and conclusion of each angel conversation. The prayers in the *De heptarchia mystica* may have had a role in the religious life of the world to come, but Dee's translation of the calls has been lost, and neither their contents nor their ultimate purpose is known.

The angelic "keys," the *48 Claves angelicae*, resemble apocalyptic

129. Whitby, *A WS* I:143–144. This volume is now British Library Sloane MS 3189.
130. For these prayers, see Dee, *A WS* II:258–302.
131. Whitby, *A WS* I:144.
132. This volume is now British Library Sloane MS 3191, f. 32r–51r. An altered version appears in a modern edition as John Dee, *The Enochian Evocation of Dr. John Dee*, ed. and trans. Geoffrey James (Gillette, NJ: Heptangle Books, 1988), pp. 17–64.
133. Whitby, *A WS* I:142–143.   134. Ibid., p. 143.

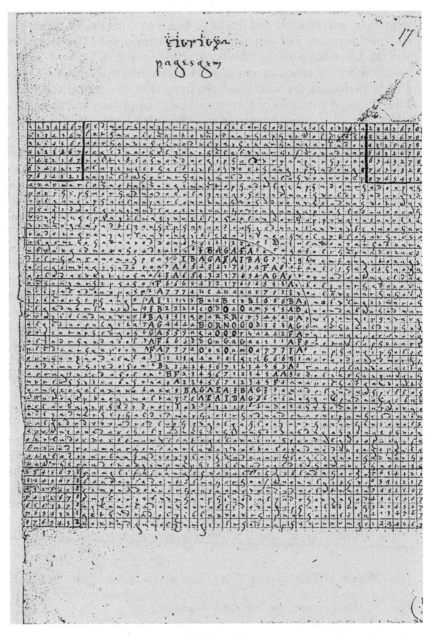

Figure 4. A page from the *Liber Logaeth* showing numbers and letters inscribed in a table. The central circle contains the name of one of Dee's angels, Bornogo. British Library, Sloane MS 3189, f. 17ʳ. Reproduced by permission of the British Library.

biblical passages in tone and content. They are remarkably similar to this example:

> Behold sayeth your God, I am a Circle on Whose hands stand 12 Kingdomes: Six are the seats of living breath: the rest are as sharp sickles: or the horns of death wherein the Creatures of ye earth are to are not [*sic*] Except myne own hand which slipe and shall ryse. In the first I made you Stuards and placed you in seats 12 of government giving unto every one of you powre successively over 456 the true ages of tyme to the intent that from ye highest vessells and the Corners of your governments, you might work my powre: powring downe the forces of life and encrease, continually on the earth. Thus you are beco[m]me the skirts of Justice and Truth. In the Name of the same your God Lift up, I say, your selves. Behold his mercies floresh, and [his] Name is beco[m]me mighty amongst us. In whome we say I move, Descend and apply your selves unto us as unto the partakers of the secret wisdome of your Creation.[135]

Here the status of the angels' revelations as a new scriptural authority is clear, and the revealed books' stylistic relationship to the Bible is evident. The *48 Claves angelicae* contained verses in both English and the angels' language, and Dee translated the keys with the angels' assistance. His finished copy of the work contained the keys in both languages.[136]

Information of a different kind concerning the angelic governors of various geographical areas and their true names could be found in the *Liber scientiae auxilii*.[137] The angels delivered this volume and its companion, the *Tabula bonorum angelorum*, which consisted of tables of letters and numbers much like those found in the *Liber Logaeth*, to Dee in central Europe after 1584. Though the books which Dee used in the angel conversations, when compared to the extant early modern grimoires and books of experiments, are more striking in their differences than in their similarities, the *Liber scientiae auxilii* is most reminiscent of traditional occult practices due to the formulaic nature of its contents.[138] Names of particular angels were plugged into standardized prayers, much like the invocations common in early modern books of magical experiments, and these prayers do not have the tone of scriptural authority struck in the *48 Claves angelicae*.

135. British Library, Sloane MS 3191, f. 4ʳ.
136. The *48 Claves angelicae* are now part of British Library Sloane MS 3191. A modern edition of this text is contained, with some alterations in format, in Dee, *The Enochian Evocation of Dr. John Dee*, pp. 65–102.
137. Whitby, AWS I: 142.
138. The *Liber Scientia* is now British Library MS Sloane 3191, ff. 14ʳ–31ʳ and 52ʳ–80ᵛ. See also Dee, *The Enochian Evocation*, pp. 103–177.

In addition to these surviving books, there were other revealed books destroyed by angelic command in Prague in 1586. During an episode when the angels claimed that all the earth had to be cleansed in preparation for the unfolding of God's divine plan, they instructed Dee to place the revealed books and his angel diaries in his oratory furnace. Later, Dee rediscovered many of his books in the garden outside his house (apparently returned by angelic means), but a few were missing, including forty-eight companion books to the *48 Claves angelicae*, an English translation of the *48 Claves*, and a short book Dee referred to as the "Mystery of Mysteries and the Holy of Holies." Dee had not yet deciphered this last work, which he believed "contained the profoundest mysteries of God Himself and of the Almighty Divine Trinity that any creature will ever live to know."[139] The loss of such a valuable work caused Dee great distress, although he took it stoically and remained convinced that, at some future date, he would understand its contents even if it was no longer in his possession.

Another book from Dee's library collection, the *Book of Soyga*, has been connected to his angels even though it was not revealed by them. The book was first mentioned in Dee's personal diary on 17 January 1582 and in the angel diaries on 10 March 1582, when Dee asked the angels, "ys my boke, of Soyga, of any excellency?"[140] The angel Uriel replied that it was a book that "God's good angels revealed to Adam in Paradise."[141] Dee was absorbed in interpreting his "Book of Soyga" and asked the angels for help, but Uriel demurred, reporting that he could not help Dee understand the text any further because the angel Michael was its "interpreter." The book was discredited by the angels and was lost or stolen around 18 April 1583. Decades later, on 19 December 1595, Dee recovered the work. The manuscript resurfaced in the seventeenth century when Elias Ashmole reported that "the Duke of Lauderdale hath a folio MS. which was Dr. Dee's with the words in the first page: 'Aldaraia sive Soyga vocor.' "[142]

The long-lost *Book of Soyga* became, in modern scholarship on Dee, a "missing link" that had vanished along with any hope of really understanding the remaining angel conversations.[143] The recent discovery of

139. Dee in Josten, p. 249.
140. The reference to the "Book of Soyga" was omitted from Halliwell's edition of Dee's diary. See Bodleian Library Ashmole MS 487, entry for January 1582. In the angel diaries, see Dee, *AWS* II:17.
141. "Liber ille, erat Ada[m]ae in Paradiso revelatus, per Angelos dei bonos." Dee, *AWS* II:17.
142. Whitby traces the history of this book in Whitby, *AWS* I:146–147.
143. See, for example, ibid.

two copies of it has not shed any further light on the angel conversations, however.[144] Its Latin and English contents appear to belie Dee's description of the book as "Arabik." The work is astrological in focus, though it does contain tables of letters and numbers – much like the revealed *Liber Logaeth*.[145] One provocative similarity between the *Book of Soyga* and Dee's angel conversations, for example, are the tables of repeating letters that resemble, in some cases exactly, the tables of letters revealed by the angels in their mystical books.[146] Nor do ownership marks permit its certain placement in Dee's collection, but marginalia and provenance suggest that the copy now in the British Library belonged to Dee, while the copy now at the Bodleian Library was copied from Dee's manuscript when he sent the *Book of Soyga* to Oxford for expert advice.[147] To add to the confusion, the *Book of Soyga's* antecedents are puzzling, as is Dee's intense interest in the work and his conviction that it was somehow connected to his angel conversations. Further study of the *Book of Soyga* may yet yield insights into the angel conversations and Dee's views on astrology and magic.

144. I located two copies of the work, one at the Bodleian Library and one in the Sloane Collection of the British Library: Bodleian Library Bodley MS 908 and British Library MS Sloane 8. I am indebted to Julian Roberts and Andrew Watson, who generously took the time to discuss the works with me after their discovery.

145. Dee, *AWS* II: 331–332.

146. For this information, I am indebted to Jim Reed who generously shared with me his analysis of the tables' structure and their relationship to the contents of Sloane 3189. See J. Reeds, "John Dee and the Magic Tables in the *Book of Soyga*," in Stephen Clucas, ed., *John Dee: Interdisciplinary Studies in Renaissance Thought* (Dordrecht: Kluwer Academic Press, forthcoming).

147. The British Library copy probably passed from Lauderdale's hands into those of Sloane. The Bodleian Library copy (which may have been copied from Dee's in the late sixteenth century by a member of Oxford University) was received from William Dun, along with four other manuscripts, in 1605. See William Macray, *Annals of the Bodleian Library Oxford* (Oxford: The Bodleian Library, 1984), p. 422. Dun may have been the "Mr. William Dunne ... doctor of phisick and late felowe of Exeter College" to whom William Marshall left his medical books. See *The History of the University of Oxford*, ed. James McConica (Oxford: Clarendon Press, 1986), 3:241. Further references to Dunne can be found in *Alumni Oxoniensis* I, Boase's *Register of Exeter*, p. 46, and Munk's *Roll of the College of Royal Physicians*, 1:102. For more information, see Roberts and Watson, *John Dee's Library Catalogue: Additions and Corrections* (London: Bibliographical Society, 1995).

The Angels

Dee, his scryers, and the many objects they used represent only part of the story of the angel conversations – though admittedly a central and vivid part. Though Dee's scryers and their sordid reputations eclipse the other features of the angel conversations, he did believe that their revelations came from the mouths of angels. Dee took the angels seriously and obeyed their commands to the letter – even when the tasks the angels assigned to him, such as denouncing the Holy Roman emperor for his sins, lacked courtly sensitivity. The angels, oddly enough, have been almost completely ignored by scholars, yet without some introduction to their individual and collective attributes the full significance of the angel conversations remains elusive.

Dee communicated with a dizzying array of angels during his conversations, both the well known and the obscure, all of whom possessed individual personalities, appearances, and demeanors. Though the majority of the angels appeared as men, a few angels important to the evolution of the conversations appeared as women. Dee's tendency not to concern himself with the angels' fixed places in the cosmic hierarchy further complicates the situation. Rather than focusing on their discrete place in the universal system, which might help us to understand how they coincide with those of other traditions, he took a more cabalistic approach and focused on their proper names, and on the particular task each fulfilled in the natural world. To the modern mind, so far removed from a worldview where celestial hierarchies of angels mediate between humanity and divinity, Dee's interest in contacting such a wide variety of angels may seem bizarre, and his complete willingness to perform any task they set before him even more inexplicable.

The most pressing question for most modern scholars remains that of Dee's intentions in the angel conversations. Did he intend to speak only with *angels*, God's divine messengers, or would he have been content to contact lesser spirits or demons? The question is complicated by Dee's interchangeable use of the words "angel" and "spirit," a common usage in his time. Cardano, for example, explained that both terms shared a common linguistic root: "the Greeks are wont to call these [beings] angels, nevertheless the Latin word is *spiritus*."[148] Dee not only used "angel" and "spirit" for his divine messengers; he also used the Aristotelian category of "intelligences," which he believed were actually "the

---

148. Girolamo Cardano, *The Book of My Life (De vita propria liber)*, trans. Jean Stoner (London: J. M. Dent & Sons, 1931), p. 240.

angels of God."[149] While it is tempting to consider Dee's angels lesser spirits – especially those whose names are unfamiliar – he believed they were divinely sent messengers, with the exception of "wicked or illuding spirits" that occasionally crept into the conversations and were quickly recognized and dismissed.

Dee's celestial communicants, whether referred to as "spirits," "intelligences," or "angels," were important, divine members of the cosmic structure. The best-known angels in Dee's conversations are the archangels of the Judeo-Christian tradition: Gabriel, Michael, Raphael, and Uriel. Their names, unlike many others which appear in the conversations, should be recognizable to anyone familiar with the Bible, with the possible exception of Uriel, who appears mainly in the now apocryphal book of Esdras. Their primary function was to serve as messengers between God, other members of the celestial hierarchy, and humanity. In addition, Gabriel, Michael, Raphael, and Uriel, according to Dee and his predecessor Agrippa, presided over the four corners of the heavens and thus had a panascopic view of the created world.[150]

Gabriel first appeared in the angel conversations on 22 June 1583.[151] The angel of annunciation, resurrection, and revelation, he guarded Eden and would remain vigilant at the entrance to Paradise until the restitution of the natural world.[152] Some of Dee's other authorities, such as Trithemius and Azalus, regarded Gabriel as the guardian of the moon.[153] Agrippa further elaborated upon the correspondences between

149. The passage appears in Dee's copy of Pompilius Azalus's *De omnibus rebus naturalibus* (Venice, 1544), p. 17ʳ. The passage reads: "Caelum enim nec sensum habet, nec intellectum, nec aliquam congitionem, nec voluntatem, *nec a seipso movet[ur], sed ab Angelis Dei, quos Aristoteles intelligentias vocat*, volvitur, ac in suos fines regulatur" (Dee's emphasis).

150. Dee, *AWS* II:7; Agrippa, *DOP*, p. 24.   151.   *TFR*, p. 24.

152. Gustav Davidson, *A Dictionary of Angels, Including the Fallen Angels* (New York: Macmillan, 1967), pp. 117–119. Davidson's work, though written in a popular style, is nonetheless a valuable reference tool that draws upon a wide variety of sources (theological, occult, and traditional). Davidson always refers to his sources, though not often as precisely as one might wish.

153. Azalus, *De omnibus rebus naturalibus*, p. 17ʳ; Johannes Trithemius, *Steganographia*, ed. Adam McLean, trans. Fiona Tait, Christopher Upton, and J. W. H. Walden (Edinburgh: Magnum Opus Hermetic Sourceworks, 1982), p. 96. Dee was so fascinated by Trithemius's *Steganographia* that in 1563 he wrote to William Cecil from Antwerp and asked to prolong his stay there so that he could finish copying the book from a manuscript in circulation in the city. See J. E. Bailey's edition of the letter in *Notes*

Gabriel and the moon by linking the angel to the mineral crystal and the metal silver.[154] Dee's understanding of the etymology of Gabriel's name was "the growing power or mightiness, or increasing strength of God."[155] A prominent figure in Islamic as well as Christian angelology, Gabriel was responsible for dictating the Koran to Mohammed and had an established connection to holy books and writing. The Jewish mystical tradition of the cabala, for instance, often featured Gabriel as the angel who delivered the cabala to humanity because of his identification with the "man clothed in linen" bearing a writing case, in Ezekiel 9:2–11. As the angel of the Annunciation, Gabriel's function as messenger to the Virgin Mary (Luke 1:26) occupied an equally important place in the history of Christianity. Early modern authors were especially intrigued by the angel Gabriel, perhaps because in Trithemius's *De septem secundeis* the historical period overseen by him began in A.D. 1525; Dee and his contemporaries were thus living in the age of Gabriel.[156] Gabriel also fulfilled a healing function in the angel conversations, embodying a "medicine" sent by God to heal Dee's wife, Jane, a role commonly associated with Raphael in orthodox Judeo-Christian traditions.[157]

Michael also features prominently in Dee's angel conversations. In the Jewish, Christian, and Islamic traditions the archangel Michael occupied a focal point in religious life and literature.[158] Guardian of Jacob, guiding angel for the nation of Israel, and the chief angelic agent in the apocalypse of Saint John the Divine, Michael was frequently represented by Christian artists. His chief attributes were his scales, which he would use to weigh human souls at the Last Judgment, and his sword of justice, which related to the meaning of his name: "strength of God."[159] According to Trithemius, Michael was the guardian of the sun, while in

and Queries, 5th ser., xi (1879): 401–402 and 422–423. Dee's copy has not been recovered; Roberts and Watson, #DM165.

154. Agrippa, *DOP;* p. 121.   155. Dee, *AWS* II:7.

156. Dee's three copies of Johannes Trithemius, *De septem secundeis* (Frankfurt, 1545) are Roberts and Watson #678, #969, and #1884. The first copy was part of the traveling library Dee took to the Continent; the second and third copies remained in England. Though both #678 and #1884 have been recovered, the first is of limited use despite its heavy annotations due to the extensive cropping of pages for rebinding; the second bears no ownership marks and so cannot be definitely attributed to Dee, although Roberts and Watson believe that his ownership is likely.

157. *TFR*, p. 250.   158. Davidson, *Dictionary of Angels*, pp. 193–195.

159. Dee, *AWS* II:7.

Agrippa, Michael was guardian of the metal quicksilver.[160] Both authors gave him important alchemical connotations by equating him with this metal and the planet Mercury. Dee's experience of the angel Michael fit neatly into this traditional framework. Michael first appeared, holding his sword of justice, on 14 March 1582.[161] At that time he told Dee that he would reveal secrets of the Book of Nature that he had not revealed to Solomon.

The angel Raphael was introduced into the angel conversations at an early point in the relationship between Dee and Kelly on 15 March 1582, just five days after Kelly's first scrying session.[162] He was not as frequent a guide as Gabriel and Michael but was the only angel known to have participated in the conversations that took place late in Dee's life. Raphael, whose name can be translated as the "healing power of God," is typically accorded the role of the angel of medicine. In Jewish mysticism, the *Zôhar* charged Raphael with healing the earth and humanity.[163] Raphael was also the angel of knowledge in most angelologies, which stemmed from his position as the guardian of the Tree of Life in Paradise.[164] In the Jewish and Christian traditions, for example, Raphael guided Tobit's son on his journey from Nineveh to Media and revealed many secrets about the future of the earth and its peoples.[165] One of Raphael's key revelations in Dee's angel conversations was the "true cabala of nature," which related specifically to his role as the angel of knowledge.[166] Knowledge of the future was also often associated with Raphael, and in both Trithemius and Azalus he was one of the seven angels of the apocalypse.[167]

Another, lesser-known archangel, Uriel, had a particularly important role in the surviving angel conversations. He was the first angel to speak with Dee during Kelly's tenure as scryer and shared many prophecies and detailed, alchemical parables with them. In the course of the conversations, Uriel explained the terrestrial significance of the angels' revela-

160. Trithemius, *Steganographia*, p. 96; Agrippa, *DOP*; p. 121.
161. Dee, *AWS* II:28.   162. Ibid., p. 36.
163. Davidson, *Dictionary of Angels*, p. 240. For more information on the place of the *Zôhar* in Christian cabala, see François Secret, *Le Zôhar chez les kabbalistes chrétiens de la Renaissance*, Etudes Juives, vol. 10 (Paris: Mouton, 1964), passim.
164. Davidson, *Dictionary of Angels*, p. 240.   165. Ibid.
166. *TFR*, p. 77.
167. Trithemius, *Steganographia*, p. 96; Azalus, *De ominibus rebus naturabilis*, p. 17^r.

tions and solved problems and answered questions that cropped up during the revelations. Dee believed that the etymology of "Uriel" was the "light of God," and Uriel obligingly clarified the derivation of his name in their conversations.[168] Though Uriel appeared as the angelic guide to Esdras in the Bible, he was more prominent in the Jewish tradition; most of the treatises mentioning him were extracanonical in the Christian West by Dee's lifetime. In cabalistic lore, Uriel transmitted knowledge of the cabala to humanity, as well as the art of alchemy, which might explain why he conveyed alchemical parables to Dee.[169] Trithemius mentioned Uriel in his *Steganographia*, where Uriel was called a "great prince" and played an important role in his system of angelic communication.[170]

The angel Anael appeared in the only conversation to survive from Dee's relationship with Barnabas Saul, but conversations with this archangel might have been extensive. Anael was traditionally one of the seven angels of creation, having guardianship over various kings and kingdoms on earth, and he was popular in occult texts of the medieval and early modern period.[171] In Dee's conversation with Anael the archangel confessed that he had power over all things, good and evil, and guardianship over the planet Venus. Etymologically, Dee believed that Anael's name denoted "the favored, wretched misery of God."[172] Dee considered the archangel Anael the "Chief governor Generall of this great [historical] period, as I have Noted in my boke of Famous and rich Discoveries," which differed from the scholarly consensus that the early modern period was under the governorship of Gabriel.[173]

The mighty and relatively well-known archangels were not the only ones to appear in Dee's angel conversations; there was also a host of other angels who appeared nowhere else in Jewish, Christian, or Islamic angelologies. Because of their idiosyncracy and mischievous natures, Calder concluded that they were fairies or daemons.[174] But Dee considered the other spiritual beings who appeared in his showstone as members of the lowest and most numerous order of the celestial hierarchy: the angels. Angels, according to the Christian authority Dionysius the Areopagite, were "more concerned with revelation and . . . closer to the

---

168. Dee, *AWS* II:7. For other variant meanings for Uriel's name, see Davidson, *Dictionary of Angels*, pp. 298–299.
169. Davidson, p. 298.   170. Trithemius, *Steganographia*, p. 84.
171. Davidson, p. 17.   172. Dee, *AWS* II:7.   173. Ibid., p. 15.
174. I. R. F. Calder, "John Dee Studied as an English Neoplatonist" (diss., The Warbing Institute, London University, 1956), 1:761.

world" than the higher orders of archangels.[175] This description was supported by the offices the lesser angels held in Dee's conversations. The angel Salamian, for example, was the angel of the sun in Dee's system. Och, the angel of Dee's "direction" or purpose in life, appeared as the guardian of the philosopher's stone and alchemy in other early modern occult systems.[176] Other angels, such as the angel Nalvage (who helped Dee with the angelic language) and the angel Ilemese (who revealed several prayers to Dee), were frequent participants in the angel conversations.

Dee's Audience

Although the majority of Dee's angels were unknown to his contemporaries, the same cannot be said of the illustrious patrons and participants who comprised their immediate audience. Because the divine plan for the terrestrial levels of the cosmos must reflect the same ordered hierarchies as the celestial levels of the cosmos, the angels told Dee to share his angelic revelations with some of the most powerful people in early modern Europe. Dee's conversations with angels had never been entirely private, and in the city of Prague they became public and, according to some observers, dangerous. Lutheran Budovec's assertion that Dee was prophesying among the people of the city has been impossible to substantiate, but if he was accurate, then Dee would have been seen as threatening, not only because of the content of his messages, but also because he was delivering them to a public audience without elite mediation. It is clear from Budovec's remarks that news of the angels and their messages had escaped from the control of aristocrats with an immediate knowledge of the conversations. Finally, Dee had caught the attention of a wider audience.

Dee's angel conversations were meant to be shared so that the "worthy" could fulfill God's intentions and purpose. "We were expressly instructed," he wrote, "from the very first beginning of that our vocation and function, and we have known ever since, that it is in accordance with our duty and most agreeable to the Divine Majesty to show those mysteries in passing, to relate them compendiously, or to give a very brief account of an action . . . to the worthy. . . ."[177] The prescribed au-

---

175. Dionysius, "Celestial Hierarchy," in *Pseudo-Dionysius: The Complete Works* (New York: Paulist Press, 1987), p. 170.
176. Dee, *AWS* II:35 and 234; Davidson, *Dictionary of Angels*, p. 211.
177. Dee in Josten, p. 226.

dience for Dee's angel conversations included members of the nobility, rulers, and religious officials both in England and abroad. A smaller subset of people actually witnessed the conversations, and their reactions ranged from skepticism to enthusiastic endorsement.

Dee's noble patrons were especially important to the perpetuation of the angel conversations, for without them he could not have gained the attention of rulers or religious officials. One of the first admitted to the colloquium of angels was Adrian Gilbert, the well-connected half-brother of Sir Walter Raleigh. Gilbert, like Raleigh, was interested in navigation and first became associated with Dee during efforts to chart the Northwest Passage with the British navigator John Davis. The two men were mentioned in Dee's private diary on 18 October 1579.[178] At that time Gilbert was entangled in a dispute with William Emery, who most believe was Dee's first scryer.[179] The first mention of Gilbert in the angel diaries appears on 23 March 1583, when the angels recognized him as an important third member of the colloquium.[180] In the angels' revelations, Gilbert was chosen to fulfill a missionary role among "the infidels" prior to the apocalyptic restitution of the world.[181] Gilbert and Kelly were openly antagonistic, however, and even Dee was plagued with doubts about Gilbert's appropriateness for such an important task, given his worldliness and irresponsibility.[182] Despite these difficulties, Gilbert remained a frequent visitor to the Dee household at Mortlake, and he might have participated in other angel conversations for which no written record now exists.

Gilbert's privileged role in the conversations did not last, and he was replaced by a Polish visitor to England, Count Albert Laski (1536–1605). The Palatine of Sieradz, Laski was close to both John Dee and Edward Kelly, a relationship possibly fostered by his interest in alchemy.[183] Described by Evans as a "strong though unorthodox Catholic," his entree into elite English circles was facilitated by visits his relative, the Protestant reformer Johannes à Lasko, made to England between 1548 and 1553.[184] Count Laski first arrived in England in the

---

178. See Roberts and Watson, p. 47.

179. See Dee, *PD*, p. 6; Whitby, *AWS* I:53–54; Roberts and Watson, p. 47. Emery would have been a teenager on this date and this would have fit the traditional age restrictions of a scryer.

180. Dee, *AWS* II:220–237.   181. Ibid., p. 240.   182. Ibid., p. 250.

183. See Evans, *Rudolf II and His World*, pp. 219–221. Laski's biography has been drawn from Evans's account.

184. Alastair Hamilton, *The Family of Love* (Cambridge: James Clarke and Company, 1981), p. 32.

spring of 1583, when Elizabeth I entertained him with all the ceremony due an important visiting nobleman, including visits to the University of Oxford to hear disputations in which another European natural philosopher, Giordano Bruno, was participating.[185]

Laski was first introduced to Dee on 18 March 1583 through correspondence, and the two men were able to meet personally on 13 May 1583 in the rooms of the earl of Leicester at Greenwich.[186] The association between Dee and Laski was cordial from the beginning, and on 18 May 1583 Laski visited Dee at Mortlake.[187] Laski honored Dee with a more extensive visit on 15 June, accompanied by "Lord Russell, Sir Philip Sydney, and other gentlemen."[188] After the party left Mortlake, Dee spoke with the angels, who mentioned Laski in their prophecies for the first time.[189] The count visited Mortlake again on 19 June 1583, when he stayed overnight and took part in his first angel conversation.[190] Many of the angels' remarks to Laski predicted his political triumph over his king, Stephen of Poland, who later also participated in the angel conversations.

The final meeting between Dee, Laski, and Kelly prior to embarking for the Continent took place on 2 August 1583. In September, the group traveled to Laski's estates in Poland, where they remained until August 1584, when Dee and Kelly departed for Prague after the angels instructed them to tell the Holy Roman emperor, Rudolf II, about their revelations. Laski remained in close contact with Dee and the angels throughout their stay in central Europe, although he was not as highly esteemed by 1586 as he had been in 1583. In 1586 and 1587 the angels cast doubts on Laski's faith in their revelations, despite his ongoing financial support of Dee and Kelly.[191]

In Prague, Don Guillén de San Clemente, the Spanish ambassador, acted as an intermediary between Dee and Rudolf II. The ambassador provided Dee with his formal introduction to the emperor in September 1584 and facilitated their initial exchange of letters.[192] San Clemente was sympathetic to Dee's interests and confided that he was descended from the Spanish mystical philosopher Ramon Lull. Such a confession might

---

185. Further connections between Bruno and Dee have remained conjectural. See Frances A. Yates, "Giordano Bruno's Conflict with Oxford," *Journal of the Warburg and Courtauld Institutes* 2 (1939): 227–242; John Bossy, *Giordano Bruno and the Embassy Affair* (New Haven and London: Yale University Press, 1991), pp. 23–27.

186. Dee, *PD*, pp. 19–20.   187. Ibid., p. 20.   188. Ibid.

189. *TFR*, p. 17.   190. Dee, *PD*, p. 20; *TFR*, pp. 22–23.

191. Whitby, *AWS* I:33.   192. Clulee, p. 223.

have elicited Dee's trust, and he showed the ambassador the fourth book of the angel diaries on 25 September 1584. San Clemente did not participate in any of the angel conversations which survive, but he was in contact with Dee throughout his time in Prague. The ambassador even took part in the christening of Dee's son Michael in 1585.[193]

Another nobleman, Vilem Rozmberk, was elevated to an important position in the plan for the restitution of the natural world in the autumn of 1586, following Laski's fall from favor. A patron of alchemists in central Europe second only to Rudolf II, Rozmberk was a natural choice for inclusion in the angel conversations as they became increasingly charged with alchemical overtones.[194] He sponsored a number of alchemists, including Heinrich Khunrath of Leipzig (1560–1605), whom Dee met on 6 June 1589 in Bremen.[195] The angels requested Rozmberk's presence at an angel conversation on 1 May 1586, but the first extant conversation involving Rozmberk is dated 14 October 1586.[196] At the time, Rozmberk was acting as peacemaker between Rudolf II and Dee, but his efforts were in vain. The contents of the angel conversations involving Rozmberk were primarily concerned with the angels' knowledge of alchemy, and they lend further credence to the belief that the rift between Dee and the emperor stemmed from alchemical disputes and difficulties rather than from the angels.

Many of Europe's rulers were also concerned with Dee's angelic revelations. Queen Elizabeth I apparently knew of his efforts to communicate with the angels, though there is no evidence that she participated in any conversations. Dee recorded on 10 March 1575 that the queen "willed [me] to fetch my glass so famous, and to show unto her some of the properties of it, which I did."[197] As no scryer was present, Dee must have exhibited only the stone's superficial, optical properties. Later, after his return from central Europe, Dee delivered a "he[a]venly admonition," which we can assume was angelically revealed, to the queen on 3 May 1594 in the privy garden at Greenwich, which the queen received "thankfully."[198] The details of their conversation were not noted.

---

193. Evans, *Rudolf II and His World*, pp. 222–223.   194. Ibid., p. 212.

195. Dee, *PD*, p. 31; Halliwell misconstrues the name as "Kenrich Khanradt." See also Evans, *Rudolf II and His World*, pp. 213–214.

196. *TFR*, p. 445.

197. Dee, *CR*, p. 17. Whitby does not believe that this glass could have been one of the angelic showstones, but was instead a mirror (see Whitby, *AWS* I:140). The reference is not sufficiently clear to rule out the possibility that Elizabeth saw one of the showstones.

198. Dee, *PD*, p. 49.

In Europe, the highest-ranking individual associated with the angel conversations was Rudolf II, the Holy Roman emperor. Dee publicly expressed his own interest in the Hapsburgs in 1564 when he dedicated the *Monas hieroglyphica* to Rudolf's grandfather. Rudolf II's deep interest in the spiritual, mystical, and occult properties of nature would have given Dee confidence that he would be welcome at the Emperor's court.[199] At first, Dee and Kelly enjoyed a positive reception, probably due to Dee's alchemical reputation rather than his conversations with angels. Once in Prague, he waited over a month before requesting an audience from Rudolf II. On 3 September 1584 Dee's request was granted, and during their interview he rebuked the emperor for his sins and demanded that Rudolf put total faith in the angelic revelations.[200] "If you will hear me, and believe me, you shall Triumph," Dee told Rudolf, but "If you will not hear me, The Lord, the God that made Heaven and Earth . . . will throw you headlong down from your seat." Few monarchs would have suffered such remarks with good grace, and Rudolf II declined to take part in the conversations.

Another monarch, Stephen of Poland, was introduced to the angel conversations on 27 May 1585.[201] The old enmity between Laski and his king was put aside, and additional audiences followed. King Stephen, unlike Rudolf II, did agree to communicate personally with the angels, though he was not predisposed to believe in the conversations, as he understood that direct revelations were unlikely now that corruption and disorder reigned in the world.[202] While there is some evidence to suggest that Dee shared Stephen's belief that magic and miracles were less likely to occur in their troubled times than in the past, he argued strongly for the divinity of his angelic revelations.[203] Dee made three specific points in his disputation. First, while conceding that the race of

---

199. See Evans, *Rudolf II and His World*, passim, especially pp. 196–242.
200. *TFR*, pp. 230–231. 201. Ibid., pp. 404–406.
202. Ibid., pp. 404–408.
203. Dee's rationale might have derived from his reading of Avicenna; Roberts and Watson, #395, p. 100ʳ. In Dee's copy of Avicenna's *De almahad*, he underlined a passage regarding the decline in spiritual activity due to increases in understanding and wisdom: "Ex quibus patet, qualiter homines vexen*tur a demon*ibus, & *a spiritibus domesticis praedictis* quia secundum hanc opinion*em animae ignorantiu[m] & sapientium vitios*orum qui in hoc mundo tantum fuerunt occupati circa sensibilia mundana si fuerint bonae, aut malae retine[n]tur in hoc mundo inferiori, ut dictum est, & continuantur cum hominibus." In the margins Dee wrote: "This agreeth wᵗʰ the walking of spirits that hath been wᵗʰin this 100 yere and before when people wer more ignorant and simple, &tc."

prophets who had prepared for Christ's birth was extinct, he argued that new prophets and prophecies had arisen to take their place. Second, nothing in the angel conversations went against God: the conversations were a special type of revelation that would prevail until the world adopted a single faith. Dee's third and final point was that twenty-four works in Greek, Latin, and English supported the angel conversations, as well as the revealed books that he intended to submit to the king's scrutiny. Despite these assurances and his participation, Stephen failed to express serious interest in the angel conversations.

On a higher spiritual if not social level, Dee and Kelly were in contact with Catholic clergy, from bishops and papal nuncios to Jesuit confessors. Dee visited with several Catholic clerics in central Europe, including the Capuchin monk Annibale Rosselli, who published a multi-volume commentary on the Hermetic *Pimander*.[204] Kelly confessed to a Jesuit professor of theology in Prague in April 1586, who took advantage of the opportunity to question him about the angels.[205] The Jesuits in Prague appear to have been particularly intimate with Dee's household and with the details of the church's case against the natural philosopher and his angel conversations. It was their interest, as well as the interest of the papal nuncios, which caused Dee such anxiety.

Dee's serious troubles with the Roman Catholic church began in 1584 when Giovanni Francesco Bonomi, the bishop of Vercelli and apostolic nuncio, prepared to leave Prague. At that time Dee came into possession of a fragment of a letter sent from the bishop to Rudolf II that included details of a dinner conversation which cast doubt on his conversations with angels. Though the substance of the document was based in "various rumours," the contents were still damaging to Dee's reputation. Dee's difficulties with Rome stemmed from his offer to share the angel conversations with Rudolf II "of his own accord and without having made a request to those who have power and authority to determine [the verity of] an apparition of blessed spirits."[206] These sentiments were echoed by Kelly's Jesuit confessor, who urged the scryer to submit the "matter of your revelations" to an "authority higher than mine."[207]

Far from arguing that Dee's angel conversations were fraudulent, the

204. Frances Yates, *Giordano Bruno and the Hermetic Tradition* (London: Routledge and Kegan Paul, 1964), pp. 179–180 and 188. Dee also received communion at the religious house where Rosselli was installed, though Yates overstates the case when she writes: "This shows us where lay Dee's true spiritual home – in religious Hermetism."

205. Dee in Josten, pp. 234–237.   206. Ibid., p. 228.

207. Ibid., p. 237.

church was concerned that the activities of Dee and Kelly were all too real, but might involve evil spirits rather than good angels. The church's position was based on the presumption that the "apparition of good angels . . . does not happen in a distinct shape which is perceived by human eyes, but is somehow vaguely encompassed . . . while . . . in a state of ecstasy and rapture."[208] The Catholic church's skepticism mounted because "Dr. Dee had a wife and was thus given to the cares of this life and to worldly matters." Therefore, the authorities argued, "it would hardly be possible for him to enjoy the intercourse of good angels, for that happened only to very holy persons, living far away from their married quarters, and to solitary hermits."[209] Of equal – and perhaps greater – concern to the church were Dee's alchemical experiments, which cast further doubt on Dee's intentions and reliability. "I am indeed of the opinion," wrote the Bishop of Vercelli, "that they prefer one philosopher's stone to ten visions of angels."[210]

Germanus Malaspina, the bishop of San Severo, assumed responsibility for the inquiries after the bishop of Vercelli left office in July 1585. Dee and Malaspina were introduced by a nobleman in Rudolf's court who told Dee that the bishop was eager to meet and befriend him. Dee was suspicious and believed that Malaspina "was preparing violence and laying in ambush for me."[211] According to Dee, Malaspina's requests persisted for more than eight months until, in late March 1586, gracious invitations turned into threatening summons. After agonizing over the matter with Kelly, Dee and his scryer decided to call on the bishop on 27 March 1586.[212] The audience opened with the bishop deploring the spread of heresy and the fragile safety of the see of Rome, which was only preserved by the king of Spain. Given the dangers of the time, the papacy had decided that "various revelations, illuminations, and consolations from the good angels of God and from God himself . . . are private, not public . . . [counsels of] reformation."[213] This was both a veiled warning that their messages might be tolerated if kept to themselves and a strategem to cast doubts on Dee's claim that he and Kelly were required by God to share their revelations with others.

Malaspina went on to ask Dee and Kelly for help with "those evils affecting us all" since the Reformation.[214] Dee wisely demurred, saying that it was not his place to give advice on such great matters, especially as he had not "received . . . any express advice of God or admonition

---

208. Ibid., p. 228.   209. Ibid., p. 229.   210. Ibid., p. 229.
211. Ibid., p. 230.   212. Ibid., p. 231.   213. Ibid., pp. 231–232.
214. Ibid., p. 232.

from the angels" to do so.[215] Though Dee admitted that "the very great and very many mysteries and counsels of God are known to us . . . which all human talents conjoined could not invent," they were nonetheless under a divine "curb of silence."[216] Dee explained that he and Kelly led a virtually monastic life, and only with reluctance "let such manifest evidence" of their angel conversations be shown.[217] Kelly was not so circumspect. After pointing out the excesses and faults of Rome, Kelly went on to voice his hope that "the doctors, shepherds, and prelates mend their ways; may they teach and live Christ by their word as well as by their conduct."[218] If these changes were made, Kelly assured the papal representatives, "a great and conspicuous reformation of the Christian religion would be brought about most speedily."[219]

Not surprisingly, the Catholic authorities were suspicious, and Francesco Pucci (1540–1593?), one of Malaspina's associates, was drawn into the angel conversations. Pucci began participating in the conversations on 6 August 1585.[220] Though he was not an enthusiastic member of the Catholic church when his contact with Dee and the angels began, an angelic message later prompted his reversion to Catholicism.[221] A lax Catholic known for his unorthodox religious beliefs, Pucci was not the most effective liaison between Dee and his critics; his close relationship with the mystic Christian Francken had already caused concern.[222] In fact, Pucci became one of the most destructive influences on the angel conversations, in part because of his role in the prophesied reformation as a "speaker" or communicator of the angel conversations to those lacking an immediate experience of the event.[223]

Nevertheless, the relationship among Dee, Kelly, and Pucci was productive until Pucci came under the influence of a third papal nuncio in Prague, Filippo Sega, the bishop of Piacenza. The bishop became alarmed at the behavior of the Dee–Kelly circle soon after he took

215. Ibid.   216. Ibid.   217. Ibid.   218. Ibid., p. 233.

219. Ibid.   220. *TFR*, p. 409.

221. Ibid., pp. 413–417. Pucci's attempts to be reconciled to the church are recorded in his letters to "Madre Lisabetta Giambonelli" and "Fratello Giovanni." See Francesco Pucci, *Lettere, documenti e testimonianze*, ed. Luigi Firpo and Renato Piattoli, Opuscoli Filosofici Testi e Documenti inediti o rari no. 11, 2 vols. (Florence: Leo S. Olschki, 1955–1959), 1: 67–69.

222. Evans, *Rudolf II and His World*, p. 225. Pucci eventually was seized by the Inquisition in 1592 and burnt for heresy. Calder, "John Dee," 1:824.

223. Dee in Josten, p. 247.

office in 1586.[224] In May 1586 Sega invited Dee and Kelly to travel to Rome, where the angels' revelations could be examined by the Catholic authorities. Pucci supported the bishop, and Dee and Kelly quickly deduced that Pucci's loyalties had switched from the contents of the angelic revelations to the doctrines of the Catholic church. Dee declined to go to Rome, as he believed that "all pious and genuine sons of the Catholic Church . . . will own [up to] the efficacy and the diaphoretic celestial energy of our Illuminator and Comforter" and would "be enriched by the knowledge" of the angel conversations.[225] This reply incensed the bishop, who accused Dee of necromancy and recommended that he be expelled from the city of Prague.[226]

In 1586, Dee's angel conversations began to stagger under the burden of too many patrons and suspicious eyes, too many angels and their individual personalities. The stresses placed on upon them reveal much about the complicated mechanics of the conversations. But the events of 1586 are not an ideal focal point to further our understanding of why Dee believed so strongly in the importance of the angels' messages. To delve deeper into the angel conversations and Dee's participation in them we must turn away from the last days of the Dee–Kelly collaboration and return to the origins of Dee's interest in communicating with angels, the roots of which can be found in a time long before Kelly became a part of the Dee household. In England, within Dee's library and his published works, we can find the first evidence that he was coming to terms with the problems of practicing natural philosophy in a world that he believed was coming to an end. While doing so, Dee began to seek a means to bridge the immense gap between the human and the divine through conversations with angels.

224. The bishop of Piacenza wrote, on 29 April 1586: "Giovanni Dii et . . . suo compagno sono in questa corte buon pezzo fa, et vanno camino di farsi autori d'una nuova superstitione, per non dire heresia, sono noti all'imperatore et a tutta la corte." Quoted in Evans, *Radolf II and His World*, p. 223.
225. Dee in Josten, p. 227.   226. *TFR*, p. 424.

2

Building Jacob's Ladder
The Genesis of the Angel Conversations

... and behold a ladder set up on the earth, and the top of it reached
to heaven: and behold the angels of God ascending and descending
on it.

—Genesis 28:12

One of the most striking biblical images is Jacob's ladder as revealed to
him in a dream at Bethel. This passage promised that communication
between heaven and earth was possible, articulated the links between the
supercelestial and the terrestrial levels of the cosmos, and suggested that
communication between heaven and earth could be reciprocal. The question that faced natural philosophers interested in bridging the gap between celestial and terrestrial was how to build and ascend such a ladder.
John Dee, along with many of his contemporaries, searched a variety of
authoritative treatises for information on how to ascend "Jacob's ladder" to learn the secrets of the cosmos, and then descend to share that
information with a waiting world. Dee's library furnished him with the
tools and materials that served as a foundation for his conversations
with angels, and they provide us with the intellectual context for what
he sought to accomplish.

Building a ladder that linked heaven and earth promised a natural
philosopher like Dee the intellectual attainment of certain knowledge, as
well as the moral achievement of spiritual salvation. Dee had been striv-
ing for both long before he became involved in his angel conversations.
From his youth Dee attended lectures at the finest universities in Europe,
discussed natural philosophy and theology with English and European
scholars of high repute, and collected the greatest library in England. He
established networks of patronage to support his work and family, and
regularly, though not prolifically, published original treatises on natural
philosophy. By 1581, however, none of these activities had provided him
with the certain knowledge of the natural world or the moral salvation
vital to his success in natural philosophy.

Dee was frustrated by this impasse, and he was not alone. Evidence from Dee's library and marginalia, as well as the angel diaries, suggests that his personal intellectual situation was linked to the Renaissance humanistic crisis of intellectual authority that questioned the reliability of centuries of accumulated wisdom.[1] If the intellectual traditions on which most scholars relied were not to be trusted, as the humanists suggested, where could a natural philosopher like Dee search for reliable information? Dee continued to use his books as a starting point for his investigation of nature, but time and again his eyes were drawn upward to the immutable celestial spheres for answers to his questions about the natural world and its complex workings. In the process, Dee was influenced by esoteric scholarship regarding the natural world and its promise that divinely sanctioned natural philosophers could ascend to the highest levels of wisdom and understanding.

Two rich resources informed these developments: Dee's library and the intellectual exchange he enjoyed with scholars and philosophers throughout Europe. By the time Dee was engaged in the angel conversations he owned the largest private library in England.[2] The library was not only sizable, its holdings were also extensive: an examination of his library catalogue reveals that he possessed a wider range of texts than could be found in any other private, university, or college library in the country. Hebrew language texts, mathematics, magic, and alchemy jostled for primacy on Dee's bookshelves. The variety and richness of the collection are matched only by the extensive connections Dee had with other natural philosophers and intellectuals at universities and glittering European courts. Dee's studies began in earnest under the tutelage of John Cheke and the Cambridge humanists, from whom he learned Greek and rhetorical skills, and received an introduction to mathematics.[3] On

1. For the intellectual crisis of authority in the early modern period, see Richard H. Popkin, *The History of Scepticism from Erasmus to Descartes* (New York: Harper, 1968); Charles G. Nauert, *Agrippa and the Crisis of Renaissance Thought* (Urbana: University of Illinois Press, 1965); Stephen A. McKnight, *The Modern Age and the Recovery of Ancient Wisdom: A Reconsideration of Historical Consciousness 1450–1650* (Columbia: University of Missouri Press, 1991).

2. For a detailed analysis of Dee's library, see Roberts and Watson, passim. For the importance of Dee's library to intellectual life in England, see William H. Sherman, *John Dee: The Politics of Reading and Writing in the English Renaissance* (Amherst and Boston: University of Massachusetts Press, 1995).

3. See Deborah E. Harkness, "The Scientific Reformation: John Dee and the Restitution of Nature (Ph.D. diss., University of California, Davis), pp. 82–92. For the ways in which mathematics was incorporated into university

the Continent, Gerard Mercator and Gemma Frisius helped Dee to master the basics of navigation, astrology, and astronomy.[4] No stranger to the patronage opportunities and intrigues of royal courts, Dee spent time in Italy with the duke of Urbino and spoke with Ulisse Aldrovandi in Bologna.[5]

It was in Paris from 1550 to 1551, however, that Dee became familiar with proponents of universal science and religion such as Jean Fernel, Antoine Mizauld, and Guillaume Postel.[6] Many universal sciences focused on a single field of natural philosophy, such as mathematics or mnemonics, but some sought to combine a variety of natural philosophies into a single science with a unified theoretical basis. Because attempts to craft a universal science diverged widely in substance, their common intentions are often buried under intricate and seemingly unrelated layers of discredited early modern systems of thought. Each universal science did possess, however, one common goal: to bridge the perceived gap between the imperfect, sublunar world and flawed human understanding and the perfect, celestial world and complete knowledge. A universal science, therefore, was thought of as an *intermediary* – a vehicle that could move its practitioner between heaven and earth. Dee's most cherished intellectual goal as a natural philosopher was to master a universal science so that he could climb to heaven like the angels in Jacob's vision at Bethel, achieve certain knowledge, and return to communicate his knowledge to others. Dee's commitment to a universal science can be detected as early as 1558, when the first edition of the *Propaedeumata aphoristica* was published. As time passed, his synthetic intellect became more intrigued with the possibility of a universal sci-

educations in the period, consult Mordechai Feingold, *The Mathematicians' Apprenticeship: Science, Universities, and Society in England, 1560–1640* (Cambridge: Cambridge University Press, 1984).

4. Harkness, "The Scientific Reformation," pp. 93–118.
5. Ibid., p. 135. For Aldrovandi, see Dee's annotations in Conrad Gesner, *Bibliotheca universalis* (Basel, 1574), Roberts and Watson #282, p. 681: "Huic dum esse [m] Bononiae (aº 1563) pollicebar chama[e]leantis et halie top historiam viva[m]." I am indebted to Julian Roberts for bringing this volume to my attention, now Bodleian Library Arch. H. c. 7.
6. On universal sciences and their place in Western European intellectual developments, see Stewart C. Easton, *Roger Bacon and His Search for a Universal Science* (Oxford: Basil Blackwell), and Klaus Vondung, "Millenarianism, Hermeticism, and the Search for a Universal Science," in *Science, Pseudoscience and Utopianism in Early Modern Thought*, ed. Stephen A. McKnight (Columbia: University of Missouri Press, 1992); pp. 118–140.

ence, which he explored in both the *Monas hieroglyphica* (1564) and the "Mathematical Preface" (1570).

Dee's conception of a universal science grew more inclusive over the years, and by 1580 he was casting his net widely to embrace optics, mathematics, cabala, angelology, eschatology, and alchemy. This seems like a bewildering variety, but if we recontextualize Dee and his work with insights from his library and from the work of his European colleagues, Dee's varied intellectual interests seem neither strange nor exceptional. A number of scholars who influenced him, including Johannes Reuchlin, Tritheimius, Henry Cornelius Agrippa, Paracelsus, Guillaume Postel, Antoine Mizauld, Jean Fernel, and Cornelius Gemma were exploring an equally wide range of possible strategies for universal sciences. In addition, all were interested in both the manifest and the occult properties of the natural world, and all believed it was possible for a natural philosopher to understand and harness the powers inherent in that world. Their common caveat was that nature could be mastered and utilized only in accordance with the divine plan and God's approval.

God's original, ordered plan for the cosmos stood behind all of these natural philosophers and their universal sciences, lending coherence to what would otherwise have been chaos. The enormous range of ideas expressed in the universal sciences stemmed not from confusion but from a lack of consensus on how natural philosophers might gain access to God's divine blueprint for the created world. For Antoine Mizauld, as for Dee, the heavens offered one of the best avenues for a comprehensive, universal natural philosophy. For Reuchlin, it was the divine word of God, decoded with the assistance of a Christianized cabala. Guillaume Postel believed that a restoration of a single religion, under a single authoritative figure, with a common body of knowledge and a common language, would enable mankind to recover the divine plan and craft a universal science. Dee explored all of these theories, and many others, in the years leading up to his decision to converse with the angels about God's created world and the natural philosophy that would disclose its secrets.

Dee's decision to search for a universal science was not sudden but developed from his earlier studies in natural philosophy. We can delineate this development through book purchases, his published natural philosophical works, annotations in his personal diary and the margins of his books, and his increasing preoccupation with the occult properties of nature and the interpretation of prophetic signs embedded in the natural world. Dee's published work often yields the clearest sense of his developing interest in universal sciences and illuminates his gradual adoption of the angel conversations as the most certain basis for a truly comprehensive natural philosophy. Topics discussed in the angel conversations – the divine language, mathematics, optics, universal science, the interpretation of natural and supernatural signs, and the apocalyptic future of humanity and the created world – all appeared in Dee's published works before they surfaced in the angel diaries, suggesting a development of his views rather than an abrupt transition between earlier "scientific" and later "magical" periods.

This chapter will argue that Dee journeyed in a step-by-step fashion toward his angel conversations. First, he lay the foundations for the conversations by studying two authoritative texts: the Book of Scripture and the Book of Nature. Second, Dee set out to observe the Book of Nature, which he explained in his *Propaedeumata aphoristica* (1558/ 1568). Observation proved an insufficient method for grasping the complexities of the created world, so Dee sought a way to decipher the Book of Nature in his *Monas hieroglyphica* (1564). Once the Book of Nature had been deciphered, however, Dee found that it was disordered and of limited use to a natural philosopher, so he relied on mathematics as a reordering principle, in his "Mathematicall Preface" (1570).

Laying the Foundations: The Books of Scripture and Nature

The foundations of Dee's universal science lie not in an arcane philosophical text but in the fundamental and widely held early modern belief that God expressed his will through three aspects of the creation. Often referred to as "books," these three manifestations of God's creative power were: the "book" of the human soul, through which salvation could be attained and which would be read at the Last Judgment; the Bible, or God's wisdom as it was revealed in the Book of Scripture; and the created world, or Book of Nature.[7] Each book, scholars believed, offered insights into both the mind of God and the act of creation. While the book of the soul, constrained by the human tendency toward sin,

7. Margreta de Grazia, "The Secularization of Language in the Seventeenth Century," *Journal of the History of Ideas* 41 (1980): 319–329, p. 319. For the Book of Nature, see Hans Blumenberg's *Die Lesbarkeit der Welt*, 2 vols. (Frankfurt am Main: Suhrkamp, 1981). For the Book of Scripture, see Beryl Smalley, *The Study of the Bible in the Middle Ages* (Oxford: Basil Blackwell, 1983); and Christopher Hill, *The English Bible and the Seventeenth-Century Revolution* (London: Penguin, 1993).

continued to preoccupy theologians, others turned their attention to the Bible and the Book of Nature for answers to pressing questions about the world and its future. The Books of Nature and Scripture became essential texts for early modern natural philosophers like Dee because they were "written" by God for humanity's benefit, and because they were accessible through study and contemplation.[8] During the fifteenth and sixteenth centuries, however, the Books of Nature and Scripture became increasingly competitive in the struggle to establish intellectual authority.

For Dee, the Book of Nature was an especially important manifestation of God's divinity. Citing Paul's Epistle to the Romans, Dee wrote that the world had no excuse not to praise God's goodness, wisdom, and power "even if it had no written memorial of these [truths other] than that which from the Creation has been inscribed by God's own finger on all creatures."[9] Dee's views predate Galileo's assertion in *The Assayer* that the Book of Nature was written in the language of mathematics with characters that were geometric in form.[10] While this passage has become emblematic of the interest in nature and observation in the seventeenth century, the metaphor can be traced back to the Middle Ages and was certainly present in Dee's work. Ernst Curtius closely studied the evolution of the metaphor, confirming that "the concept of the world of nature as a 'book' originated in pulpit eloquence, was then adopted by medieval mystico-philosophical speculation, and finally passed into common usage."[11] Despite Dee's hearty endorsement of the Book of Nature as an authoritative text, and the belief of Galileo that the language of that text was mathematical, not even by the time of Isaac Newton had the Book of Nature's authority eclipsed that of the Book of Scripture.[12]

In practice, both the Book of Scripture and the Book of Nature were problematic texts. Replete with symbolism, allegories, analogies, and

8. Dee, MH, p. 125.   9. Ibid.
10. Quoted in de Grazia, "Secularization of Language," p. 320. For more on Galileo and the Book of Nature, see Joseph C. Pitt, *Galileo, Human Knowledge, and the Book of Nature* (Boston: Kluwer Academic Publishers, 1992); Eileen Reeves, "Augustine and Galileo on Reading the Heavens," *Journal of the History of Ideas* 52 (1991): 563–579.
11. Ernst Robert Curtius, *European Literature and the Latin Middle Ages*, trans. Willard R. Trask, Bollingen Series vol. 36 (Princeton, NJ: Princeton University Press, 1973), p. 321.
12. See B. J. T. Dobbs, *The Janus Faces of Genius: The Role of Alchemy in Newton's Thought* (Cambridge: Cambridge University Press, 1991), pp. 57–66.

inconsistencies, the two books presented an early modern natural philosopher with a serious obstacle: how were they to be "read"? A critical,
humanistic stance already had been taken toward the Bible, but the Book
of Nature presented additional problems. What language, natural philosophers wondered, had God used to inscribe the Book of Nature? What
was the ordering principle of the text? Was the Book of Nature stable,
or had it deteriorated since the Fall?

Theologians like Saint Cyprian had been grappling with these issues
for centuries before humanistically trained natural philosophers turned
their attention to them. In his "Epistle to Demetrianus" Cyprian wrote:

> the world has now grown old, and does not abide in that strength
> in which it formerly stood; nor has it that vigour and force which it
> formerly possessed. This, even were we silent, and if we alleged no
> proofs from the sacred Scriptures and from the divine declarations,
> the world itself is now announcing, and bearing witness to its de-
> cline by the testimony of its failing estate.[13]

In Saint Cyprian and other early Christian authors the concept of a
decaying world blended Hellenistic notions with ideas from the Old
Testament, the Apocrypha, and the Pseudepigrapha.[14] Biblical passages
from Isaiah, the Psalms, and 2 Esdras are especially relevant to the
development of these ideas about natural deterioration through their
references to the Book of Nature's aging, or "wearing out," process. In
Isaiah, the aging process was not limited to the earth but included the
heavens: "The earth is mourning, withering, the world is pining, withering, the heavens are pining away with the earth" (Isaiah 24:4).

By Dee's time ancient and medieval authors' ideas about the instability
of the Book of Nature were used by humanistic scholars to suggest that
the created world was a text (like many others) in need of careful
reading, rereading, and restitution.[15] Though the cosmos had been or-

---

13. Saint Cyprian, "Epistle to Demetrianus," quoted in Ernest Lee, *Millennium
    and Utopia: A Study in the Background of the Idea of Progress* (Berkeley:
    University of California Press, 1949), Tuveson, p. 13. Dee owned a 1500
    Paris edition of Saint Cyprian's works and sermons; Roberts and Watson,
    #1373.

14. See David Brooks, "The Idea of the Decay of the World in the Old Testa-
    ment, the Apocrypha, and the Pseudepigrapha," in *The Light of Nature*,
    ed. J. D. North and J. J. Roche (Dordrecht: Martinus Nijhoff Publishers,
    1985); pp. 383–404, passim.

15. See James J. Bono, *The Word of God and the Languages of Man* (Madison:
    University of Wisconsin Press, 1995), pp. 26–84, for an insightful overview
    of this process.

derly and the Book of Nature had been reliable at the time of the Creation, since that time both had declined. In the Christian West, the decay of the Book of Nature was related to the moral fall of humanity when Adam and Eve sinned and the perfect order of Paradise was disrupted. While in Paradise, Adam and Eve lived in an abundant garden, communicated directly with God, and led a life of ease and plenty. The Adamic lifestyle, characterized by a harmonious relationship between God, humanity, and nature, was God's intended plan. Once outside the garden, humanity's struggle to master the Book of Nature began, and all subsequent changes represented a deterioration from the Edenic ideal.[16]

While the details of the moral fall were peculiar to the Judeo-Christian tradition, the notion of a paradise or idyllic garden that provided a reference point of order and stability for the Book of Nature was not. Many ancient cosmologies and cosmogonies contained references to a time in the past when humanity, divinity, and nature had lived in harmony and concord. In pagan antiquity, this idyllic state was commonly referred to as the "Golden Age," a time of perfection that underwent a series of subsequent degenerations.[17] Early Christian authors combined pagan accounts of the Golden Age with biblical ideas of Paradise, creating a conception of past perfection more complex than the discrete traditions from which it was made. Philo Judaeus, for example, believed that the perfection of the Edenic past was embodied in Adam himself. In Philo's works, Adam took on the proportions of a giant who was physically perfect and free from disease as well as intellectually gifted. Because Adam's intellect was linked to the divine, he was able to grasp "the natures, essences, and operations which exist in heaven," rather than being fettered by earthly restrictions.[18] Augustine held a similar

16. Frank L. Borchardt, *Doomsday Speculation as a Strategy of Persuasion: A Study of Apocalypticism as Rhetoric*, Studies in Comparative Religion, vol. 4 (Lewiston, NY: Edwin Mellen Press, 1990), pp. 45–57.

17. John Prest, *The Garden of Eden: The Botanic Garden and the Re-Creation of Paradise* (New Haven and London: Yale University Press, 1981), pp. 18–19.

18. F. R. Tennant, *The Sources of the Doctrines of the Fall and Original Sin*, introduction by Mary Frances Thelen (New York: Schocken, 1968), pp. 137 and 149. Dee owned four works by Philo Judaeus: his *Antiquitatum libri* (Basel, 1527), Roberts and Watson, #101; the complete *Opera* (Basel, 1561), Roberts and Watson, #164; *Libri quatuor* (Antwerp, 1553), Roberts and Watson, #334; and *De mundo* (Basel, 1533), Roberts and Watson, #1052.

belief: that Adam was immortal and "had special knowledge which empowered him to give the animals the correct names."[19]

The expulsion of Adam and Eve from Paradise was not the only event to destabilize the Book of Nature. Of equal importance was a second disruption: the Flood. For sixteenth-century scholars and their predecessors the Flood hastened nature's deterioration and erased the last vestiges of Paradise from the world. In addition, the deluge of the waters and their receding caused physical changes within the Book of Nature. Land that had been flat and easily traversed became interrupted by mountains and valleys after the moral Fall and Flood. Earthquakes and storms had not afflicted Paradise, but emerged with the decay of the Book of Nature. Some believed that changing seasons and climatic shifts, including excessive rain and drought, were actually a result of the Flood.[20] One of Dee's contemporaries, Francis Shakelton, asked his readers to observe the Flood's all-too-visible consequences: "doe we not see the yearth to be changed and corrupted? Sometymes by the inundation of waters? Sometymes by fiers? And by the heate of the Sunne?"[21]

As Shakelton suggests, Dee's contemporaries were anxious about the disorder and instability of the Book of Nature. The chaos was so pronounced that many thought it could only culminate in the end of the world, as foretold in the Book of Scripture. English clergyman Hugh Latimer, in a sermon of 1552, pointed to the natural signs of deterioration and change – including rings around the sun and eclipses – "which no doubt signifies . . . that this fearful day is not far off."[22] Such tensions only increased when, in 1572, a new star appeared in the heavens – the first new star since the star of Bethlehem that announced the birth of Christ.[23] A few years later, in November 1577, a comet blazed through the heavens.[24] After the comet appeared, Elizabeth I called on Dee to

---

19. Augustine, *Enchiridion*, p. 239; quoted and discussed in Borchardt, *Doomsday Speculation*, p. 47.

20. Prest, *The Garden of Eden*, pp. 16–17.

21. Francis Shakelton, *A blazyng Starre* (London, 1580), sig. Aiiii.

22. Quoted in Tuveson, *Millennium and Utopia*, p. 49.

23. The best-known treatise on the subject of the new star was Tycho Brahe, *De nova stella* (1573). For an English translation, see Tycho Brahe, *His Astronomicall Coniectur of the New and Much Admired * Which Appered in the Year 1572* (London, 1632).

24. Dee noted the event, drawing a picture of the comet across the top of his personal diary entries for the month of November. See Bodleian Library Ashmole MS 487, entries for November 1577. Halliwell mistranscribed this drawing to 22 November 1580.

soothe her anxious courtiers and to interpret the comet's potentially eschatological message.[25] Then, during the Easter season of 1580, a rare earthquake rocked London early in the morning, an event which Dee noted in his personal diary.[26]

Astrologers of the period warned that another cataclysmic event loomed on the horizon, further threatening the stability of the Book of Nature: a conjunction of the superior planets Saturn and Jupiter. This was an astrological event of such rarity and importance that it was discussed in print for years before it actually took place in 1583.[27] Dee was well-versed in this literature, and by 1552 he owned Albumazar's important work on the significance of "grand conjunctions."[28] The predicted 1583 event was more than a conjunction; it also involved a shift from the "watery trigon" to the "fiery trigon." This shift is complicated but critical to our understanding of the importance Dee accorded to his angelic conversations as well as to the growing sense of decay and deterioration that people thought was affecting the Book of Nature. There were four celestial trigons, each related to the elements and signs of the zodiac. For example, the "watery trigon" was comprised of the

25. Dee, *PD*, p. 4.
26. Dee may also have been called to court to discuss this event, as on 11 April 1580 he noted that "I went to the court [and] ca[me] home again." Entries concerning the earthquake were omitted from the Halliwell edition of Dee's personal diary. See Bodleian Library Ashmole MS 487, entries for April 1580, including: "Terremotus hor .6. a meridie durabat p[er] duo min-uta. . . ."
27. The literature on the conjunction and its significance includes Margaret E. Aston, "The Fiery Trigon Conjunction: An Elizabethan Astrological Prediction," *Isis* 61 (1970): 159–187; Bernard Capp, *Astrology and the Popular Press. English Almanacs 1500–1800* (Ithaca, NY: Cornell University Press, 1979), pp. 164–179; Germana Ernst, "From the Watery Trigon to the Fiery Trigon: Celestial Signs, Prophecies and History," in Paola Zambelli's *'Astrologi hallucinati': Stars and the End of the World in Luther's Time* (New York: Walter de Gruyter, 1986), pp. 265–280; Robin Bruce Barnes, *Prophecy and Gnosis: Apocalypticism in the Wake of the Lutheran Reformation* (Stanford, CA: Stanford University Press, 1988), pp. 141–181; Laura Ackerman Smoller, *History, Prophecy, and the Stars: The Christian Astrology of Pierre d'Ailly, 1350–1420* (Princeton, NJ: Princeton University Press, 1994); Ann Geneva, *Astrology and the Seventeenth-Century Mind: William Lilly and the Language of the Stars* (Manchester: Manchester University Press, 1995), pp. 118–140.
28. Roberts and Watson, #421.

three signs linked to water (Pisces, Cancer, and Scorpio), and the "fiery trigon" was linked to Aries, Leo, and Sagittarius. The trigons influenced the earth in a repetitive cycle from fiery, to earthy, to airy, to watery. As each new trigon became influential, it caused significant changes in the terrestrial sphere.

These changes were magnified when a conjunction of the superior planets closest to God – Saturn and Jupiter – occurred as they entered the sign of the zodiac that signaled a new trigon. Thus a "great conjunction" of Saturn and Jupiter occurred every twenty years, when Jupiter and Saturn entered one of the three signs of a specific trigon. "Greater conjunctions" were much rarer and more significant events, occurring once every 240 years, when Jupiter and Saturn entered into a completely new trigon. The rarest conjunction of all was the "greatest conjunction." This occurred once in roughly a thousand years when Jupiter and Saturn reentered the fiery trigon – the initial trigon in the cycle – in the sign of Aries.[29] Historians and astrologers agreed that there had been six "greatest conjunctions" since the creation: one during the life of the prophet Enoch; another with Noah's Flood; a third when Moses had received the Ten Commandments; the fourth during the dispersal of the ten tribes from Israel; the fifth at the birth of Christ; and the sixth coinciding with the reign of Charlemagne.

Few doubted that the "greatest conjunction" due to occur in 1583 would be the harbinger of a similarly monumental event, especially since discussions of the conjunction were often reprinted with new editions of Regiomontanus's fifteenth-century astrological predictions that the world would end in 1588.[30] Astronomer Tycho Brahe suggested that the potential effects of the entry of Jupiter and Saturn into the fiery trigon would be magnified by the new star he observed in Cassiopeia in 1572. Brahe decided that these celestial signs, when combined, foretold unprecedented changes in the Book of Nature.[31] A Bohemian astrologer Cyprian Leowitz (1514?–1574) concurred and discussed the significance of the conjunction in his *De coniunctionibus magnis insignioribus superiorum planetarum* (1564).[32] "Since . . . a new trigon, which is the fiery,

---

29.　Aston, "Fiery Trigon Conjunction," pp. 161–162.

30.　Ibid., p. 160.

31.　Ibid., p. 164; Brahe, *His Astronomicall Coniectur*, pp. 16–18.

32.　Aston, "Fiery Trigon Conjunction," pp. 164–165. Dee's copy of this work is Roberts and Watson, #631; he purchased the work in the year it was published, 1564. Now Cambridge University Library R*.5.21, it is annotated by Dee; for an analysis see Evans, *Rudolf II and His World*, p. 222.

is now imminent," Leowitz wrote, "undoubtedly new worlds will follow, which will be inaugurated by sudden and violent changes, for this has happened before when one trigon ended and another began, but especially if the watery trigon is being followed by the fiery."[33] Leowitz was sure that the upcoming conjunction announced "undoubtedly . . . the second coming of the son of God and man in the majesty of his glory."[34]

Observing the Book of Nature: *The Propaedeumata Aphoristica* *(1558)*

Although the Book of Nature was an unstable and unreliable text – never more so than in the years prior to 1583 – a natural philosopher could not afford to ignore the manifold signs that God had inscribed there, but had to practice a shrewd observation of nature. This represented the first step of ascent on the ladder of knowledge that could take a natural philosopher to a certain understanding of the cosmos. Dee explored the potential of observation in his first work, the *Propaedeumata aphoristica* (1558).[35] Modern scholarship has provided two very different conceptions of the work and its place in Dee's natural philosophy, both of which offer clues regarding the development of the angel conversations. Peter French, for example, argues that the *Propaedeumata aphoristica* infused magic into all parts of the cosmic system.[36] Nicholas

Leowitz's work was popular enough to warrant a second edition, printed in London in 1573 just following Brahe's observation of the new star in Cassiopeia.

33. Quoted in Aston, "Fiery Trigon Conjunction," p. 165.

34. Cyprian Leowitz, *De coniunctionibus magnis* (1564), sig. L3ᵛ quoted in Aston, "Fiery Trigon Conjunction," p. 166.

35. Readers who wish to explore the *Propaedeumata aphoristica* more closely should consult Wayne Shumaker's translation of the work, *John Dee's Astronomy* (Berkeley: University of California Press, 1978), with an introduction and supporting materials by John Heilbron; Nicholas H. Clulee, "Astrology, Magic, and Optics: Facets of John Dee's Early Natural Philosophy," *Renaissance Quarterly* 30 (1977): 632–680; and Clulee, pp. 43–73. The work was published twice, once in 1558 and again in 1568, because of Dee's dissatisfaction with errors allegedly made by the printer. Few substantive changes were made, and we shall be considering the first edition, unless otherwise noted.

36. French, *John Dee*, pp. 93–96.

Clulee, on the other hand, emphasizes Dee's adherence to mathematical principles of optics to create an "astrological physics" which was an integral part of a cosmos governed by natural law.[37] What both scholars suggest is that at a very early point in his career Dee was seeking to unify the basis of natural philosophy (though there is debate as to whether that basis was optical or magical), the first step in establishing a universal science.

When Dee wrote the *Propaedeumata aphoristica* he was already worried about the instability of the Book of Nature and his ability to practice natural philosophy with the assistance of such an unreliable text. In the *Propaedeumata aphoristica* Dee articulated the basic division between a disorderly, terrestrial realm and an ordered, celestial realm. He described the cosmos as being divided into two material parts that corresponded to the dichotomy between disorder and order: an incorporeal, spiritual realm; and his own terrestrial world, "composed of unstable elements" subject to corruption and decay.[38]

Dee's *Propaedeumata aphoristica* evolved from such general cosmological concepts through more complicated, technical discussions of parallaxes, the behavior of light, and the specifics of astrological influence. Astrology was the root of natural philosophy during the early modern period, and its basic premises supported the dominant, Ptolemaic cosmology Dee adopted and provided theoretical mechanisms to explain processes such as growth and decay. Dee began to lay the universal, physical foundations for his conversations with angels in the *Propaedeumata aphoristica* through a discussion of the "rays" that all things, both immanent and occult, dispersed throughout the universe. Later, Dee used his crystal "showstone" to capture and magnify the rays on which angels were said to travel into the natural world. Dee's ideas on the propagation of rays form the unified physical basis for his astrology as well as the unified metaphysical basis of the angel conversations and display as well his mastery of Baconian optics and the light metaphysics of Robert Grosseteste. More important, the optical theories that Dee outlines in the *Propaedeumata aphoristica* provided a way of observing phenomena that had both spiritual and physical significance, thus bridging the enormous gap between the terrestrial and the celestial parts of the cosmic system.

Another cornerstone of Dee's cosmology as expressed in the *Propaedeumata aphoristica* was the ancient belief that nothing in the cosmos stood in isolation, but rather possessed "order, agreement, and similar

37. See Clulee, pp. 43–73.   38. Dee, *PA*, p. 125.

form with something else."[39] Dee suggested that similar and harmonious parts of the cosmos imitated each other and moved together in what we might call "occult physics."[40] In Dee's system, higher levels of the cosmos "activated" lower levels of the system in such a way that the universe could be likened to "a lyre tuned by some excellent artificer, whose strings are separate species of the universal whole."[41] It was through this kind of activation that "wonderful things can be performed truly and naturally, without violence to faith in God or injury to the Christian religion."[42] The natural philosopher who knew how to touch the strings of the great, cosmic lyre "dexterously and make them vibrate would draw forth marvelous harmonies," Dee argued.[43]

The key to such mastery, according to Dee's *Propaedeumata aphoristica*, was optics – the study of the behavior of light. A complete understanding of Dee's commitment to optics has been undermined because of the loss of several of his optical works, including 1557 and 1559 works on artistic perspective, a defense of Roger Bacon (the *Speculum unitatis*), and a treatise on burning mirrors or glasses. It is clear, however, that Dee's contemporaries knew, and relied upon, his optical ex-

39. Ibid.
40. I use the term "occult physics" to differentiate this idea of celestial *movement* from Nicholas Clulee's "astrological physics," which centers on celestial *influences* that cause changes in secondary qualities of motion and light. See Clulee, p. 40.
41. Dee, *PA*, p. 127. Dee was not the only natural philosopher to liken the world to a musical instrument whose strings, when plucked, caused "sympathetic" reverberations. For other important proponents of this image, including Robert Fludd, see Joscelyn Godwin, *Music, Magic and Mysticism: A Sourcebook* (London: Arkana, 1987).
42. Dee, *PA*, pp. 125–127.
43. Ibid., p. 127. Dee's use of the cosmic lyre brings to mind Marsilio Ficino's musical magic and medicine, as well as Ficino's belief in intermediary powers and cosmic harmony. The classic account of Ficino's philosophy is Paul Oskar Kristeller, *The Philosophy of Marsilio Ficino* (1943), trans. Virginia Conant (Gloucester, MA: P. Smith, 1964). For a specific analysis of Ficino's belief in intermediary agencies, see Michael J. B. Allen, "The Absent Angel in Ficino's Philosophy," *Journal of the History of Ideas* 36 (1975): 219–240. D. P. Walker's *Spiritual and Demonic Magic from Ficino to Campanella* (Notre Dame, IN: University of Notre Dame Press, 1975) also contains information on Ficino's spiritual philosophy. The most recent study dealing with Ficino's musical magic and medicine is Gary Tomlinson, *Music in Renaissance Magic: Towards a Historiography of Others* (Chicago: University of Chicago Press, 1992).

pertise. William Bourne reported that Dee was one of the pre-eminent optical philosophers of the period, ranking him above Thomas Digges. Bourne wrote to Cecil that "there ys dyvers in this lande, that . . . do knowe much more in these [optical matters] . . . then I: And specyally Mr Dee, and all so Mr Thomas Digges, for . . . by theyre Learninge, they have Reade from many moo [more] auctors."[44]

Dee's library catalogue provides ample evidence of the reading program to which Bourne referred. Early collection lists from 1556–1557 show that Dee purchased works from the medieval optical tradition as it descended from Robert Grosseteste and the Arabic philosophers down to Roger Bacon.[45] Dee consulted Bacon's major works and was familiar with rarer works as well, as he owned an abridged copy of Bacon's work on burning mirrors, *De speculis comburentibus*, of which there are only three other known copies.[46] Optics provided the foundation for Bacon's natural philosophy and theology as he worked toward a universal science. Bacon argued that understanding the behavior of light – God's first creation in Genesis – could lead to a complete understanding of nature.[47] Because the behavior of light could be understood with the assistance of geometry, mathematics was elevated to a central place in Bacon's practice of natural philosophy, as it had been derived from the work of Plotinus, al-Kindi, and Robert Grosseteste.[48]

In the *Propaedeumata aphoristica*, Dee applied Baconian theories about the behavior of light to the central problems of celestial influence in astrology. While some celestial influences were sensible, others were occult and "penetrate in an instant of time everything that is contained in the universe."[49] The question facing Dee was whether the observable behavior of sensible rays could shed light on the invisible behavior of occult rays. In the fourth aphorism Dee began to address this dilemma when he explained that all things with an active existence, both manifest and occult, emit rays but that they were more pronounced among things which were "incorporeal and spiritual" rather than those "corporeal and composed of unstable elements."[50]

44. British Library Landsdowne MS 121, art. 13, f. 101ᵛ.
45. Clulee, pp. 52–53.
46. See David Lindberg, *Roger Bacon's Philosophy of Nature: A Critical Edition, with English Translation, Introduction, and Notes, of De multiplicatione specierum and De speculis comburentibus* (Oxford: Clarendon Press, 1983), pp. lxxix–lxxx. Dee's copy, now Bodleian Library Ashmole MS 440, ff. 1ʳ–23ᵛ, is not a complete version of the text.
47. Ibid., p. vii.   48. Ibid., pp. vii–viii.   49. Dee, *PA*, p. 133.
50. Ibid., pp. 123–125.

Dee finally resolved the issue of the applicability of observable phenomena to occult phenomena through Bacon's contention that radiation was the root of all natural causation, and could explain both the manifest and the occult properties of light.[51] Bacon's theories of radiation thus could account *physically* for the occult influence of celestial bodies in the terrestrial world.

Following Bacon, Dee likened radiation to the "emission of species." Bacon's "species" were themselves derived from al-Kindi's "universal force," which radiated from every individual thing in the cosmos and produced a variety of effects.[52] Bacon argued, and Dee concurred, that both species and rays differed in their power to affect and cause effects.[53] Thus, the heavens could influence the earth more than the earth could affect the heavens, but the two were nonetheless linked through a web of radiation. Dee's understanding of the Baconian explanation of radiation and species was depicted on the title page of the *Propaedeumata aphoristica*. In the 1568 edition the title page was illustrated with Dee's alchemical *monas* symbol. Radiating from the center of the *monas* were six broken lines that extended to earth, water, the sun, the moon, heat, and humidity in the borders of the illustration. The rays on the title page were not directional; they did not depict either emanation from the center of the *monas* symbol or emanation from the natural world. Instead, they showed that the *monas* affected, and was affected by, natural radiation, thereby indicating the way in which nature both *received* and *emitted* radiation.

Dee was not solely interested in what we might call the physics of light; he was preoccupied as well with the metaphysics of light. His exploration of light metaphysics encouraged him to base his cosmology and astrology on the behavior of rays because of light's privileged position in the creation of the natural world. In Genesis 1:3, light was God's first creation, and in Genesis 1:14–19 God created specific "lights" in the heavens – the stars and celestial bodies – to separate day from night, to mark the seasons and years, and to shed light on the earth.[54] In the Gospel of John 1:1–9, divine light was equated with God, the Word of God, the creative principle, and the universal redemptive force. Dee could also have drawn on classical authors such as Plato and Plotinus

51. Ibid., pp. 127–129.
52. Lindberg, *Bacon's Philosophy of Nature*, p. lv.
53. Dee, *PA*, p. 125.
54. Dee used this image for a splendid frontispiece to four couplets addressed to William Cecil entitled "Primi Quatridui Mysterium," now Bodleian MS Ashmole 1789, f. 2b.

for insights into the mystical significance of light, but similar ideas were developed in the work of Robert Grosseteste.

Grosseteste's influence on Dee has not been adequately studied. Clulee accorded him a relatively minor position of influence in Dee's intellectual development, arguing that only Grosseteste's *De iride* could have been consulted prior to the completion of the *Propaedeumata aphoristica*, based on his consultation of Dee's 1556/1557 reading list.[55] This overlooks the possibility that Dee consulted works by Grosseteste sometime between 1557 and 1558, as well as Dee's ownership of several comprehensive manuscript collections of Grosseteste's works which are not dated and could be from this period. Dee owned at least sixteen works by Grosseteste, including works on astronomy, the behavior of light, mathematics, physics, devotional works, and romances.[56] Grosseteste and Dee had much in common: an interest in angels and celestial intelligences, a fascination with the works of Pseudo-Dionysius, a desire to decode natural symbols and biblical allegories, and a belief that the physics and metaphysics of light were crucial to any understanding of the cosmos and its divine plan.[57]

From Grosseteste, Dee would have been led to believe that light represented the unity of God's creation, and that the metaphysics and physics of light provided a method for understanding the way God operated in nature.[58] The extent to which Dee was willing to subject the metaphysical properties of light to physical and optical investigations can be surprising. At one point in the *Propaedeumata aphoristica*, for example, he suggested that the supracelestial primum mobile was like a concave spherical mirror that drew every created part of the cosmos into

55. See, for example, French, *John Dee*, pp. 93–96, and Clulee, pp. 53–54.
56. See James McEvoy, *The Philosophy of Robert Grosseteste* (Oxford: Clarendon Press, 1982), p. 519, and Roberts and Watson, p. 217, for a complete list of Dee's titles and their relationship to Grosseteste's corpus.
57. For detailed discussions of Grosseteste's life and philosophy, see A. C. Crombie, *Robert Grosseteste and the Origins of Experimental Science, 1100–1700* (Oxford: Clarendon Press, 1958); Gunar Freibergs, ed., *Aspectus et affectus: Essays and Editions in Grosseteste and Medieval Intellectual Life in Honor of Richard C. Dales*, with an introduction by Richard W. Southern (New York: AMS Press, 1993); Richard Southern, *Robert Grosseteste: The Growth of an English Mind in Medieval Europe* (Oxford: Clarendon Press, 1986); and McEvoy, *The Philosophy of Robert Grosseteste*, passim. For a survey of medieval attitudes toward the metaphysics of light, see James McEvoy, "The Metaphysics of Light in the Middle Ages," *Philosophical Studies* (Dublin) 26 (1979): 124–143.
58. Southern, *Robert Grosseteste*, pp. 218–219.

itself.[59] Dee never defined the primum mobile, but it was common to assign it to a place in the created world beyond the fixed stars closest to God, and to make it ultimately responsible for the movement of the eight celestial spheres. The primum mobile was not, therefore, part of the created world, and scholars considered it beyond the powers of human understanding. Yet Dee, likening the primum mobile's behavior to that of a concave spherical mirror, invited further analysis of its behavior through the application of optics. In Dee's natural philosophy, no facet of cosmology, including those that verged on theology, were exempt from an exploration of their physical behavior.

By the time Dee completed the *Propaedeumata aphoristica* he had already begun to take important steps toward the angel conversations. First, the *Propaedeumata aphoristica*'s emphasis on the ability of optics to provide a unified tool for the study of the Book of Nature – both its manifest and occult properties – foreshadows the role of the showstone in the angel conversations. Light was not only the universal natural philosophical principle; it was also the universal metaphysical principle. Light served, then, as an ontological intermediary between the divine, celestial spheres and the corrupt, sublunar world. Dee's interest in the power of such intermediaries only increased as his traditional studies yielded fewer and fewer certain results. Second, Dee had come to believe that a truly wise natural philosopher worked with nature, rather than against it, "forcing nature artfully" by means of "pyronomia," or alchemy.[60] In this description Dee recommended that the natural philosopher harness what God had already inscribed in the Book of Nature, rather than try to coerce nature or command supernatural powers. It was the job of the natural philosopher to hasten, delay, or better articulate natural processes – not to change, alter, or subvert them. These ideas were explored more fully in his next work, the *Monas hieroglyphica*.

Deciphering the Book of Nature: The Monas Hieroglyphica (1564)

While the *Propaedeumata aphoristica* offered a strategy for observing the Book of Nature and extrapolating theories about the occult from those observations, Dee's next work focused on deciphering and analyzing their significance. The method he used to approach these problems was linguistic: he wanted to find a better way to *read* God's messages in

59. Dee, *PA*, p. 135.   60. Ibid., p. 123.

the mysterious Book of Nature. Dee wrote that the "science of the alphabet contains great mysteries since He, who is the only Author of all mysteries, has compared Himself to the first and last letter" and inscribed the Book of Nature with his finger.[61] The problems natural philosophers experienced when trying to "read" the unstable Book of Nature were more pressing during Dee's lifetime because Galileo's confident assertion that the language of the Book of Nature was mathematical and the alphabet comprised of geometric figures had not been established as a commonplace. Rather, scholars debated the role of language in the Creation, the identity of the language Adam spoke in Paradise, and whether any traces of that language were still in the world. For natural philosophers like Dee, the discovery or recovery of the language of the Book of Nature was a matter of serious concern. In addition, many believed that the language of Nature might well be the most valuable element of a universal science.[62]

In the *Monas hieroglyphica* (1564) Dee made further attempts to unify the many branches of natural philosophy – optics, alchemy, astrology, cabala, mathematics, geometry, and astronomy. This time, however, he attempted to encompass their subtleties into a single glyph. Grammarians, mathematicians, geometers, musicians, astronomers, experts in optics, cabalists, physicians – even technicians called scryers who looked into crystals – would find their practices subsumed into the science of the *monas*. The *monas*'s virtues were inexhaustible in Dee's eyes, as it contained the generation of all letters, rendered abstract numbers corporeal, conveyed the mysteries of squaring the circle, mimicked the celestial harmonies, displayed the orbits of the heavenly bodies, gave the parabolic formula for the construction of a burning mirror, contained medicinal proportions, and aided the perception of visions in crystal lamines or carbuncles.[63] Despite Dee's enthusiasm, the work was an enigma to most of his contemporaries. When Dee returned to England shortly after the *Monas hieroglyphica*'s publication, he was obliged to explain it to Queen Elizabeth. Some years later Rudolf II, a student of the occult sciences and natural philosophy, confessed that the work was "too hard for his . . . capacity."[64] The intervening centuries have not served to make its contents less obscure.

61. Dee, MH, p. 125.
62. See Bono, *The Word of God and the Languages of Man*, passim; Umberto Eco, *The Search for the Perfect Language*, trans. James Fentress (Oxford: Basil Blackwell, 1995).
63. Dee, *MH*, pp. 127–137.
64. For the reference to Queen Elizabeth, see Dee, *CR*, p. 19; for Rudolf II, see *TFR*, p. 231.

An apparent contradiction in the Bible may provide some clues as to what Dee intended. Although Holy Scripture was divine revelation and should not contradict itself, passages in Genesis and Wisdom appeared irreconcilable on the subject of the language used by God to inscribe the Book of Nature. While Wisdom 11:21 attributed the ordering and creative principle of the cosmos to number, weight, and measure – in short, to mathematics – a passage in Genesis offered an alternative. In Genesis 1:3, God literally "spoke" the cosmos into being: "And God said, Let there be light: and there was light."[65] The linguistic ordering principle in Genesis emphasized the creative power of the Word of God, which was emphasized again in Psalm 148: "He spake the Word, and they were made; He commanded, and they were created." References to the power of God's Word were strengthened in Christianity through the belief that Christ was the Word of God, or Logos, made manifest for human salvation.

Given this scriptural contradiction – or, at the very least, ambiguity – it is not surprising that there was little consensus during the sixteenth century about the language God may have used to write the Book of Nature. John Dee, it is clear, did not embrace an explanation of the cosmos and its creation that was exclusively or even primarily mathematical. Instead, Dee was typical of his generation of natural philosophers, who sought to find a compromise between the mathematical creation described in Wisdom and the linguistic creation described in Genesis through a close study of ancient languages and the Jewish cabala. This course of study was promising, because so many ancient languages – Greek, Hebrew, Latin, and Chaldean – were alphanumeric. In alphanumeric languages each character represents both a phonetic sound and a numerical value, fusing language and mathematics, and suggesting that at one time mathematics and language were more closely linked. Early modern scholars tended to focus on Hebrew as the best avenue for exploring the mysteries of the Book of Nature, since it was thought that God had used a divine, uncorrupted form of Hebrew to create the cosmos, imparting an implicit mathematical order at the same time. Adam was believed to have shared this original and mathematically charged form of Hebrew with God, infusing the linguistic-mathematic sense of coherence into the created world when he gave all parts of nature their true names.[66]

65. An overview of these biblical ideas and their treatment by subsequent commentators can be found in Roy Harris and Talbot J. Taylor, *Landmarks in Linguistic Thought: The Western Tradition from Socrates to Saussure* (New York: Routledge, 1989), pp. 35–45.
66. Eco, *Search for the Perfect Language*, especially pp. 73–116.

Humanistic study of the development of language and the transmission of ancient wisdom challenged the belief that the language of the Book of Nature was some form of Hebrew. As natural philosophers mastered an ever larger number of languages, they began to argue that European national languages (such as Dutch, Welsh, and Italian), ancient Oriental languages (such as Chinese, Aramaic, and Syriac), pictorial languages like hieroglyphics, and the language of mathematical symbols were actually older than Hebrew and therefore more likely to be the original language used by God to inscribe the Book of Nature.[67] Natural philosophers' preoccupation with the deterioration of the Book of Nature, when combined with the linguistic findings of the humanists, made them wary of any contemporary language that purported to be God's own. Surely, they reasoned, if God had used a form of Hebrew to create the world, it was far removed from the early modern, or even ancient, language. How, then, could an early modern natural philosopher hope to recover, learn, and utilize the original language with its inherent mathematical and creative powers and its special relationship to the Book of Nature?

Several options were deserving of consideration. First, a people could still be speaking a language not far removed from the language God and Adam shared in Paradise. This theory gained strength as Native American populations were encountered by European explorers, and was also a feature of European interactions with the Chinese.[68] Proponents of exploration perpetuated the notion that as yet "undiscovered" peoples

---

67. Efforts to reclaim God's original language were widespread and continued into the seventeenth century, despite what Eco describes as a systemic confusion over "universal" languages and the original "perfect" language. See Eco, passim, for these later debates and their Renaissance context; Allison Coudert, "Some Theories of a Natural Language from the Renaissance to the Seventeenth Century," in *Magia Naturalis und die Entstehung der modernen Naturwissenschaften* (Wiesbaden: Franz Steiner Verlag, GMBH, 1978); pp. 56–114; James J. Bono, *The Word of God and the Languages of Man*. These efforts preceded seventeenth-century plans for the creation of a universally intelligible language, as discussed in Vivian Salmon's "Language-Planning in Seventeenth Century England: Its Contexts and Aims," in *In Memory of J. R. Firth*, ed. C. E. Bazell et al. (London: Longmans, 1966); pp. 370–397, and James Knowlton, *Universal Language Schemes in England and France 1600–1800* (Toronto: University of Toronto Press, 1976). A more specific treatment of how efforts to understand and shape the universal language impacted on science is M. M. Slaughter, *Universal Languages and Scientific Taxonomy in the Seventeenth Century* (New York: Cambridge University Press, 1982).

68. Umberto Eco, *The Search for the Perfect Language*, pp. 72–116.

might be using the divine language and even suggested that Paradise might be found and restored.[69] These scholars argued that the Garden of Eden had survived the Flood but had been isolated by the deluge and was now protected by the angels and inhabited by the saints. Should Paradise be reclaimed, they argued, vestiges of the linguistic knowledge of Adam would be recovered. As more of the earth's surface was explored, however, even enthusiastic natural philosophers like Dee grew skeptical of such claims. Dee's knowledge of terrestrial geography even led him to argue with the angels about the impossibility of locating Paradise in an angelic lesson in 1584. When his scryer described the image of "a great place, about 300 mile [sic] long, like a Park, enclosed with fire," and enthused that it must be Paradise, Dee demurred. In spite of the angel's insistence that it was indeed Paradise, Dee replied that "Till 45 degrees, both Northerly and Southerly, all is known in the most part of the world: But of any such place there is no knowledge nor likelyhood by any History of these dayes, or of old time."[70]

In contrast with explorers who sought to reclaim Paradise and thereby the divine language, some humanistically trained natural philosophers maintained hope that a pure understanding of ancient languages – usually Hebrew – might be recovered through the same careful methods that had made progress in the accuracy of Latin and Greek.[71] Natural philosophers acknowledged that, although the written form of a language might be stripped of its corruptions, it could not be restored to its original power, because there would not be a clear sense of how the language had been spoken – a major source of its potential efficacy. The significance of the spoken power of language was underscored in magical texts of the ancient and medieval periods. Michael Psellus (fl. eleventh century), for example, confirmed in his commentary on the Chaldaean oracles that there were "god-given names among all races, which have unspeakable power in magic rites."[72] He noted, however, that an act as

69. Christopher Columbus, for example, encouraged by Pierre d'Ailly's early-fifteenth-century collection of astronomical and astrological texts *Imago mundi* (1483), believed that his explorations were playing a key part in the recovery of the lost Garden of Eden. Pauline Moffitt Watts, "Prophecy and Discovery: On the Spiritual Origins of Christopher Columbus's 'Enterprise of the Indies,'" *American Historical Review* 90 (February 1985): 73–102, pp. 74–75. John Dee's copy of d'Ailly is Roberts and Watson, #272.

70. *TFR*, pp. 156–157.

71. See Thomas C. Singer, "Hieroglyphs, Real Characters, and the Idea of Natural Language in English Seventeenth Century Thought," *Journal of the History of Ideas* 50 (1989): 49–70, passim.

72. Quoted in John Dillon, *The Golden Chain: Studies in the Development of Platonism and Christianity* (Aldershot, Hampshire: Variorum, 1990), p. 211.

simple as translating powerful words from one language to another could diminish their efficacy. For this reason, Psellus warned his readers, "*Do not*, therefore, *change* . . . [Hebrew words] into the Greek language, e.g. Seraphim and Cherubim and Michael and Gabriel. For these when uttered in the Hebrew form have unspeakable power in magic rites, but if changed into Greek words, they lose it."[73] Obtaining an ancient text in its original language was difficult, and the problems posed by incorrect pronunciation or translation became more troubling still as humanists underscored the myriad ways in which language could fall victim to inaccuracies, changes, and misunderstandings.

Despite these difficulties, the divine language of Creation remained a powerful, if elusive, lure in the sixteenth century. Guillaume Postel (1510? – 1581), one of Dee's associates in Paris, believed so strongly in the efficacy and power of the divine language that he made it the cornerstone of his universal science.[74] Postel's efforts informed Dee's attempts to build a universal science on a linguistic foundation. Dee acquired additional insights into the value of ancient languages from his purchase of Postel's works, as well as an Arabic grammar, historical works, and several cosmological treatises.[75] During his visit to Paris he became more acutely aware of the enormous potential, and the enormous difficulties, associated with the study of the natural world.

Dee's *Monas hieroglyphica* presented a universal emblem – the *monas* – through which the Book of Nature could be studied and natural philosophy could be mastered. The *monas* was constructed from traditional astrological symbols infused with "immortal life," "able to express their especial meanings in any tongue and to any nation." Dee's

73. Quoted in ibid., p. 211.
74. Postel's life and thought have been the subject of two full-length studies and several shorter treatments. See especially W. J. Bouwsma, *Concordia mundi: The Career and Thought of Guillaume Postel* (Cambridge, MA: Harvard University Press, 1957), and Marion L. Kuntz, *Guillaume Postel, Prophet of the Restitution of All Things: His Life and Thought* (The Hague: Martinus Nijhoff Publishers, 1981), passim. Shorter studies include W. J. Bouwsma, "Postel and the Significance of Renaissance Cabalism," *Journal of the Warburg and Courtauld Institutes* 18 (1954): 313–332; François Secret, "Alchimie, palingénésie et métempsychose chez Guillaume Postel," *Chrysopoeia* 3 (1989): 3–60; Paolo Simoncelli, *La lingua di Adamo. Guillaume Postel tra accademici e fuoriusciti fiorentini*, Biblioteca della Rivista di Storia e Letteratura Religiosa Studi e Testi, vol. 7 (Florence: L. Olschki, 1984).
75. For a detailed discussion of his Parisian purchases, see Harkness, "The Scientific Reformation," pp. 109–117.

combination of symbols represented a return to an older, purer language he believed was used by the ancient wise men or magi.[76] This led Clulee to argue that Dee believed he had reconstructed "the original divine language of creation that stands behind all human languages" when he conceived the *monas* symbol.[77]

In part, Dee's achievements were humanistic in the *Monas hieroglyphica*: he felt he had restored the original characters of the divine language through minor modifications of contemporary astrological symbols. As he explained in the *Monas hieroglyphica*, the "external bodies" of the astrological figures were "reduced or restored to their mystical proportions" and thus attuned to their internal forms and God's original design. The fullest example that Dee gives of this restoration process and its effect appeared in his description of the "hieroglyphic" for Mercury, "the rebuilder and restorer of all astronomy . . . [who was sent to us] by our IEOVA so that we might either establish this sacred art of writing as the first founders of a new discipline, or by his counsel renew one that was entirely extinct and had been wholly wiped out from the memory of men."[78]

In Dee's *Monas hieroglyphica*, humanistic impulses combined with his interest in ancient forms of writing, as evidenced in Dee's description of his *monas* as a "hieroglyphic," which had a number of important associations in the sixteenth century.[79] Western interest in hieroglyphics escalated in 1419 when Cristoforo de' Buondelmonti discovered the *Hieroglyphics of Horapollo* on the Island of Andros and took his prize back to Florence, where humanists added the manuscript to a growing number of works that purported to explain the meaning and significance of hieroglyphics.[80] Centuries before the Rosetta Stone enabled linguists to

76. Dee, *MH*, p. 121   77. Clulee, p. 88.   78. Dee, *MH*, p. 123.
79. See E. Gombrich, *Symbolic Images: Studies in the Art of the Renaissance* (New York: Phaedon, 1972); Rudolf Wittkower, "Hieroglyphics in the Early Renaissance," in *Developments in the Early Renaissance*, ed. Bernard S. Levy (Albany: SUNY Press, 1972): 58–97.
80. For a modern edition of the text see George Boas's translation, *The Hieroglyphics of Horapollo (1950)*, with a new foreword by Anthony T. Grafton (Princeton, NJ: Princeton University Press, 1993). For the influence of this discovery on other cultural expressions of the period, see Charles Dempsey, "Renaissance Hieroglyphic Studies and Gentile Bellini's *Saint Mark Preaching at Alexandria*," in *Hermeticism and the Renaissance*, ed. I. Merkel and A. Debus (Washington, DC: Folger Shakespeare Library, 1988), pp. 342–365. See also Liselotte Dieckmann, "Renaissance Hieroglyphics," *Comparative Literature* 9 (1957): 308–321; Liselotte Dieckmann, *Hieroglyphics: The History of a Literary Symbol* (Saint Louis, MO: Washington

decode the essentially phonetic nature of the Egyptian glyphs, early modern philosophers and philologists theorized that the language was both universal and divine. When this idea was adopted by Neoplatonists such as Ficino and Pico della Mirandola, philosophical circles emphasized the ability of hieroglyphics to instantly denote an idea or a series of ideas. It was the Neoplatonic interest in hieroglyphics as a symbolic system of communication that fostered the growth of the popular emblematic tradition, where a picture could indeed speak a thousand words and pithy conceits could be hidden from the eyes of the vulgar.[81] Pierio Valeriano's *Hieroglyphica* (1556), with its self-conscious adoption and modification of hieroglyphics into erudite symbols for the learned to decode, became the authoritative early modern emblematic text.[82]

While the search for the divine language of the Book of Nature provided the context for Dee's *Monas hieroglyphica* and the Neoplatonic interest in hieroglyphics provided its symbolic references, the cabala provided the exegetical method that placed the work more firmly within the natural philosophical tradition. Though Meric Casaubon argued in the seventeenth century that the *Monas hieroglyphica* proved Dee was a "cabalistic man up to the ears," the implications of his argument have not been fully explored.[83] The *Monas hieroglyphica* demonstrates Dee's belief that cabala was the province of divinity, and that within its doctrines the mystery of a mathematical creation enacted through speech could be explored. He could not credit human powers of invention to have "established out of such mystical principles that very stupendous fabric of the Hebrew letters and the Nekudoth [vowel points]." Instead, "an afflation of the divine power" must have been present in the Hebrew language when it was created.[84]

Dee's interest in the cabala can be divided into two periods: the first prior to the making of his 1556–1557 library list; and the second coinciding with his trip to Antwerp in 1562, which culminated in the publi-

University Press, 1970); and Erik Iversen, *The Myth of Egypt and Its Hieroglyphs in European Tradition* (Copenhagen: GecCad Publishers, 1961).

81. Dempsey, "Renaissance Hieroglyphic Studies," pp. 343–347.

82. Dee's copy of a 1567 edition of this work is Roberts and Watson, #114; it has not been recovered. Other emblem books which Dee owned are Roberts and Watson, #720, #843, and #2140.

83. The variety of interpretation can be seen by consulting longer analyses of the work, including Michael T. Walton, "John Dee's *Monas hieroglyphica*: Geometrical Cabala," *Ambix* 33 (1976): 116–123; Clulee, 77–142.

84. Dee, *MH*, p. 127.

cation of the *Monas hieroglyphica*. The interest he demonstrated during the first phase was characteristic of a Christian cabalist. Though Roberts and Watson assert that he showed little interest in cabala prior to the 1560s, Dee already owned the works of Reuchlin, Agrippa, Giorgi, and Trithemius – a comprehensive library of Christian cabala for the period.[85] Reuchlin's *De verbo mirifico*, the classic work of Christian cabala, would have provided an adequate grounding in its principles and techniques, while Agrippa's *De occulta philosophia, libri tres*, Francesco Giorgi's *De harmonia mundi*, and Trithemius's *De septem secundeis* were devoted to more esoteric applications of the art. Agrippa, for example, devoted much of his second and third books to the magical possibilities of the cabala and borrowed a great deal of his information from Reuchlin in the process. Giorgi, on the other hand, constructed his enormous cabalistic work (the table of contents alone covers twenty pages) on a foundation of cosmology, numerology, and angelology, and posited that the created universe was bound together by principles of emanation, musical harmony, and angelic intervention.[86]

Between 1560 and 1562, Dee's earlier interest in the Christian cabala led him to study the Hebrew language. This occurred during another excursion to the Continent, when Dee was living with the printer Willem Silvius in the thriving city of Antwerp. A bibliophile like Dee would have reveled in the opportunity to come into contact with new works in Antwerp's richly textured intellectual community.[87] Specifically, An-

85. See Roberts and Watson, pp. 11 and 29. Roberts and Watson fail to distinguish between Hebrew books about the cabala and books of Christian cabala. Dee did not begin to purchase Hebrew texts and grammars until the 1560s, at least several years after we have evidence of his first interest in the mystical aspects of the language.

86. Thorndike, VI: 450–453. Dee's copy of Giorgi's text has not been located. Evidence that Dee was well versed in Giorgi's theories can be found in his copy of Gesner's *Epitome*, p. 535. Against Gesner's discussion of the works of Orpheus, Dee wrote: "Ex Orphei Libro De Verbo Sacro aliquot carmina citat Franciscais Georginus Venetus, in volumine suo De Harmonia Mundi, pag. 10. fol. 10."

87. Dee went to Antwerp from Louvain, where he bought texts relevant to his Hebraic studies and made his only datable purchase of a work by Paracelsus. I concur with Roberts and Watson, who believe that Dee switched from the old year to the new on 1 January rather than in the spring. This is certainly borne out by the angel diaries. French, Clulee, and all other biographers have, by assuming that Dee followed English rather than European calendar practices, lessened Dee's stay on the Continent by a full year. See, for instance, French, *John Dee*, p. 36. For Dee's copy of Paracelus, see Roberts and Watson, #1476.

twerp was home to the one of the largest communities of Jews and *conversos* in Europe, and it is quite possible that Dee received lessons in Hebrew and cabala in the city where he was to begin writing the *Monas hieroglyphica* in January 1564.

Dee's book purchases underscore the importance of this second period of his cabalistic study. Of the six books we know Dee to have purchased in 1561 only one is not concerned with Hebrew. These five titles are the first indication that his interest in cabala had exceeded Christian limitations, no matter how comprehensive the treatment or reputable the author. If we expand our consideration of his known library purchases to 1562, we discover that, of the twenty-two works purchased over a two-year period, twenty were concerned with Hebrew or the cabala. This pattern represents a pronounced change from Dee's eclectic purchasing habits in earlier years.[88] Though only a fraction of Dee's Hebrew collection has been recovered, G. Lloyd Jones has determined that his library housed the largest collection of Hebrew language materials in England during the early modern period – greater than those of either Cambridge or Oxford University, despite their professors of Hebrew, and greater than that of any single college within either university.[89]

Despite his unparalleled collection, Dee was never as competent in Hebrew as he was in Greek or Latin and later felt compelled to confess to the angels, "I am not good in the hebrue tung."[90] Dee's expertise certainly never eclipsed that of Reuchlin or Trithemius; his interest in Hebrew was not that of a Hebraic scholar but a cabalist. Dee's attempts to master Hebrew were made in an effort to come to terms with the intricacies of the cabala and to apply them to other intellectual projects such as the *Monas hieroglyphica* and, later, the angel conversations.[91] A

88. Any analysis of the library contents is hampered by imperfect preservation, yet this confluence of cabalistic works is striking when compared to previous and subsequent years. While this may be an indication less of Dee's interest than of later preservation, the percentages are striking. There had been a slight increase in alchemical texts in proportion to other works between 1558 and 1562, for example, but the rise was not as dramatic as that which occurred between 1560 and 1562 in Hebrew texts. See Harkness, "The Scientific Reformation," for a detailed discussion of Dee's earlier purchasing patterns, pp. 75–132.

89. G. Lloyd Jones, *The Discovery of Hebrew in Tudor England: A Third Language* (Manchester: Manchester University Press, 1983), p. 168.

90. Dee, *AWS* II:65.

91. Jones argues persuasively, for example, that Dee's interest initially stemmed from his "belief in the efficacy of the Hebrew alphabet for making contact with the spirits, and through them with God" (p. 176).

work we know made a deep impression on Dee during his stay in Antwerp supports this contention: Trithemius's treatise on cryptography entitled *Steganographia*. "Of this boke," Dee reported to Queen Elizabeth's secretary, William Cecil, "the one half, (with contynuall Labor and watch, the most part of x dayes) have I copyed out." Dee had been delayed in Antwerp because he was waiting for a Hungarian nobleman to finish copying the work in exchange for lessons on "some points of Science."[92] During the interim, Dee requested permission to remain abroad, because "Such Men, and such bokes are come to my knowledge . . . [regarding] the former[ly] mentioned great sciences" of cabala that he was reluctant to leave.[93] Trithemius's *Steganographia* is one of the most curious works of the early modern period. It suggested a means of secret communication which employed angels to carry messages from sender to recipient. Dee thought that the book's usefulness was "greater than the fame thereof is spred," and might have brought it to Cecil's attention so that the queen's secretary could use its methods for matters of state.

From his reading Dee came to believe that there were two cabalas: the "vulgar cabala," based on Hebrew and centuries of Jewish tradition; and the "real cabala," which was based on divine messages embedded in the Book of Nature. The "vulgar cabala," according to Dee, was fundamentally linguistic and relied on letters written and maintained by mankind. The "real cabala," which Dee claimed he discussed in "aphorisms to the Parisians" (possibly a lost work of 1562 entitled *Cabbalae Hebraicae compendiosa tabella*), was "born to us by the law of creation" and was more divine than the "vulgar cabala" other scholars studied and employed. Despite Dee's distinction, he nonetheless defended the worth of the Jewish tradition. His separation of the "real cabala," or, as he described it, the cabala "of that which is," from the "vulgar cabala," or the cabala "of that which is said," is essential to an understanding of the *Monas hieroglyphica*.[94] As the remarks above indicate, Dee did not intend to devalue any form of the cabala, but instead to apply its techniques properly to "that which is," as encrypted in his *monas*. The *monas*, when subjected to cabalistic exegesis, thus became capable of explaining the most occult sciences, and even inventing new arts and sciences valuable to humanity.

Throughout the *Monas hieroglyphica* Dee applied cabalistic exegetical techniques to aspects of the Book of Nature and his own universal

92. John E. Bailey, "Dee and Trithemius's "Steganography," *Notes and Queries*, (5th ser., 11, 1879): 401–402 and 422–423; see p. 402.
93. Ibid., p. 402.   94. Dee, *MH*, p. 135.

symbol. He used "tzyruph" [*zeruf* or *temurah*], "gematria," and "notar-iacon" [*notarikon*] to reveal and explore their hidden significance.[95] In *gematria*, the numerical values of Hebrew letters and words were calculated to reveal both hidden meanings within words and connections to other words with the same numerical value. In the *Monas hieroglyphica*, *gematria* was used to explicate the central cross representing the Roman numeral ten, two Roman numeral fives, and other more abstract numbers.[96] *Zeruf* or *temurah* involved the rearrangement of the letters within Hebrew words to form other words. This was the element of cabala which Dee used most heavily in the *Monas hieroglyphica*, combining and recombining the point, arc, and line of his *monas* to form astrological and alchemical symbols, Greek letters, alchemical vessels, and numbers.

Dee's most intriguing use of traditional cabalistic techniques involved *notarikon* – the expansion of a single Hebrew letter into a word. In the *Monas hieroglyphica*, he used the technique to suggest that components of the *monas* were abbreviations of other things found in the natural world.[97] Dee had used notariacal principles in the *Propaedeumata aphoristica*'s aphorism XVIII, when he referred to *notarikon* in connection with alchemical proportions, or "appropriate weights." These notariacal alchemical proportions, according to Dee, yielded insights into the creation of metals:

> In the four separate great wombs of the larger world [the four elements] there are three distinct parts; these are ... condensed, structured and regulated by their own appropriate weights and may now be called, by notariacal designation, AOS, or OSA, or SOA (for pyrologians will understand me if I speak so).[98]

This aphorism has remained inscrutable, in large part because the "notariacal designations" could not be located in the typical cabalistic references of the period.[99] Yet Dee was an uncommon alchemist as well as an unusual cabalist, as is demonstrated both in the *Monas hieroglyphica* and also in his 1559 acquisition of Giovanni Pantheus's *Voarchadumia contra alchimiam* (1530), where he discovered an approach to alchemical transformation that united cabalistic theories with elemental doctrines, and which provides the reference to the notariacal designations that feature in the *Propaedeumata aphoristica*.[100]

95. Clulee, pp. 92–95.   96. Dee, *MH*, pp. 169–173.
97. Clulee, p. 93.   98. Dee, *PA*, p. 129.
99. Wayne Shumaker, in his notes to the *Propaedeumata aphoristica*, described this aphorism as "the most inscrutable of all the aphorisms" and was unable to locate the source for Dee's ideas. See Dee, *PA*, pp. 210–213.
100. References to AOS appear scattered throughout Pantheus, *Voarchadumia* (Roberts and Watson, #D16), but their first appearance is on p. 18,

Dee's early exposure to the ideas of Giovanni Pantheus influenced his cabala, his alchemy, and also the *Propaedeumata aphoristica* and *Monas hieroglyphica*. Dee's copy of the work is heavily annotated – so heavily that blank pages were bound into the text – and he inscribed it repeatedly with his *monas*. The interleaved pages are not, however, Dee's notes. They are textual comparisons between the *Voarchadumia contra alchimiam* and Pantheus's earlier work, the *Ars transmutationis metallicae* (1518). Dee owned a 1550 Paris edition of both works, yet he still made a detailed and permanent comparison of the two treatises for his library.[101] Because the comparative, transcribed passages appear exactly adjacent to the relative printed text, the notes from the *Ars transmutationis* must have been written on pages interleaved with the *Voarchadumia* after the text was bound. This careful work indicates that Dee had a serious and sustained interest in a cabalistic alchemy of "that which is."

The similarities between the work of Pantheus and Dee are worthy of further consideration. The most important similarity for understanding Dee's *Monas hieroglyphica* and the angel conversations lies in Pantheus's contention that a "cabala of metals" could be traced back to the Chaldeans and his assertion that the cabala of metals provided a means to restore alchemy.[102] These are strikingly similar to the reforming aspirations of Dee's hieroglyphic *monas*.[103] Pantheus's notion of a "cabala of metals" is analogous to Dee's "real cabala," or the "cabala of that which is." In addition, whereas Pantheus's cabalistic manipulations were based on the fundamental, elemental level of creation ordered through numerical proportions and signified by letters of the alphabet, in Dee's work the science of alchemy was a mortal and terrestrial attempt to mimic the

where "A" designates "Materia prima artis," "O" designates "Lux maior," and "S" designates both "Ignis. i. Lux maior" and "Aer .i. Lux minor."

101. Roberts and Watson, #1437.

102. For the reference to the Chaldeans, see Pantheus, *Voarchadumia*, pp. 9–10: "Quandoquidem hac in nostra tempestate studiosorum solertia in lucem venerit ac publicam noticia[m] Chaldaea professio: qua diebus nostris facile patuit: quod antehac preosq[ue] latuit artis .s. purificationis Auri durarum caementationum perfectarum Inquisitores post ipsum (ut aiunt) Tubalcha'in primum ipsius Authorem Chaldeaeos aut Indos potius extitisse." For Voarchadumia as a cabala of metals, see Pantheus, *Voarchadumia*, p. 11: ". . . Voarchadumia est ars liberalis: virtute praedita: sapeintiae ocultae: non avara: non vana: possibilis: verissima: necessarie: & p[er]sequenter perquire[n]da: quae metallorum Ca'bala nuncupatur. . . ."

103. Clulee, pp. 101–103.

divine work of creation. Dee utilized the cabalistic techniques of gematria, notarikon, and zeruf to expound upon the geometrical, mathematical, cabalistic, alchemical, and astrological significance of his *monas*.

Dee's *Monas hieroglyphica* represented a reformed universal science capable of mimicking the divine act of creation alchemically, mediating the cosmos numerically, and reading the Books of Nature and Scripture cabalistically – a seemingly comprehensive plan. Yet there were limits implicit in his universal science, which Dee alluded to in the last theorem. In theorem XXIV, he referred to the fourth chapter of Revelations, which begins with a door opening in the heavens and an angelic voice inviting John the Divine to see into the future as well as into the secrets of Heaven. This was a moment in Revelations just prior to the breaking of the seven seals and the unleashing of the four horsemen of the Apocalypse.[104] It can be argued that Dee's *Monas hieroglyphica* was his attempt to take mankind to that point in cosmic history and that level of cosmic ontology through a universal, exegetical science so that a door would open in the heavens and further marvels could be revealed.

As the *Propaedeumata aphoristica* had pointed out, however, a natural philosopher must observe before he interprets, and Dee believed that his ability to discern signs of decay in the Book of Nature was what enabled him to interpret their significance. But it was a rare natural philosopher, in Dee's opinion, who gave up the lure of earthly riches for such noble pursuits:

> the republic of letters can muster only one man out of a thousand, even of those scholars who have entirely dedicated themselves to studies of wisdom, who has intimately and thoroughly explored the explanations of the celestial influences and events [as well as] the reasons of the rise, the condition, and the decline of other things.[105]

There was another avenue to explore, however, before Dee was willing to give up traditions of human wisdom and resort to angels for assistance in this observational and interpretational work. Mathematics had become an implicit feature of his observational and exegetical works, but it was not until his "Mathematical Preface" to the works of Euclid that Dee faced the implications of a deeply mathematical natural philosophy.

104. J. A. Van Dorsten, *The Radical Arts: First Decade of an Elizabethan Renaissance* (London: Oxford University Press, 1970), p. 23. In his analysis of the *Monas hieroglyphica*, Van Dorsten writes that the work "leads but to a first degree of revelation, or initiation: the opening of a door in heaven before the seals of the book are opened."

105. Dee, *MH*, p. 117.

Ordering the Book of Nature: The "Mathematical Preface" (1570)

By the time Dee wrote the "Mathematical Preface" he was asking his readers to grasp how the mathematical sciences – rather than optics, alchemy, or the cabala – constituted the root of a universal natural philosophy and provided a reliable way to access the secrets of divinity and the intricacies of the natural world.[106] The possibility of a mathematical ordering principle offered another way for natural philosophers to circumvent the problems of decay and deterioration afflicting the Book of Nature. In his "Mathematical Preface," Dee discussed the mathematical underpinnings of several natural philosophical disciplines, including optics, astrology, navigation, and technology, providing his readers with another potentially universal science.

Although the biblical passage in Wisdom did much to perpetuate interest in the mathematical nature of the creation, the notion that the cosmos was created with numbers ultimately derives from theories attributed to the Pythagoreans. And, in important passages in both the *Republic* and the *Timaeus*, Plato echoed the Pythagorean belief that the cosmos was mathematically perfect.[107] Platonic and Pythagorean ideas were revived in the early modern period by Marsilio Ficino and other Neoplatonists. Dee was intrigued with the divinity of mathematics and the Neoplatonic conviction that the act of creation was essentially mathematic. Both ideas appeared in his preface to the first English translation of Euclid's *Elements*, where he included the Platonic argument that mathematics should be considered divine because of the mathematical

---

106. John Dee's "Preface" in Euclid, *The Elements of Geometry*, trans. Henry Billingsley (London, 1570). Dee's preface has been reprinted as *The Mathematicall Praeface to the Elements of Geometry of Euclid of Megara (1570)*, ed. Allen G. Debus (New York: Science History Publications, 1975), along with an introduction and notes. For a provocative analysis of the work, see Kenneth K. Knoespel, "The Narrative Matter of Mathematics: John Dee's Preface to the *Elements* of Euclid of Megara (1570)," *Philological Quarterly* 66 (1987): 26–46. For a comparison between the attitudes of Dee and another sixteenth-century natural philosopher toward mathematics that focuses specifically on Euclid, see Enrico I. Rambaldi, "John Dee and Federico Commandino: An English and an Italian Interpretation of Euclid during the Renaissance," *Rivista di Storia della Filosofia* 44 (1989): 211–247.

107. See Debus's "Introduction," in Dee, MP, p. 5.

nature of God's creative process.[108] Though Dee was not the only natural philosopher to explore the ramifications of these ideas in the sixteenth century, it was not until the seventeenth century that mathematics became the dominant language for discussing and describing nature.

In the opening remarks to his "Mathematical Preface" Dee argued that the importance of mathematics rested on numbers and the intermediary role they played between natural and supernatural levels of the cosmos. As intermediaries, numbers participated in everything from the divine to the mundane. Dee asked his readers to explore this mysterious duality rather than simply considering numbers and geometrical figures for their utilitarian value. For, while all numbers and mathematics shared essential traits, they were not indistinct. The highest and most abstract state of number was reserved for the Creator. A more mundane and natural state of number existed in every creature in the world. The third state of number, analogous to the intermediary, mathematical level of the cosmos, could be found "in Spirituall and Angelicall myndes, and in the Soule of ma[n]." Form and function intertwined in this level, with the angels serving as intermediary agencies between the divine and the terrestrial while occupying the intermediary level of the cosmos.

In his "Mathematical Preface" Dee referred to both "numbers numbering" that were in God and the angels and to "numbers numbered," in creatures occupying the terrestrial world. The pure numerical forms reserved for God, however, were totally immaterial, and therefore essentially distinct from the numbers that existed in the human spirit and the angels. Neither pure element nor Aristotelian quintessence, Dee stated, provided the true "matter" of these divine numbers, and not even the spiritual or angelic substance was immaterial enough to contain the numbers used by God in the Creation.[109] Dee based his arguments on traditional cosmological concepts already present in his *Propaedeumata aphoristica*, especially the ancient division of the cosmos into terrestrial and celestial levels. In the "Mathematical Preface," Dee developed this distinction, arguing that the natural world was comprised of things material, compounded, divisible, corruptible, and changeable. Thus, it was perfectly acceptable to judge things in the natural world by applying standards of probability and to reach decisions through conjecture.[110] The natural world was, in Dee's eyes, inherently uncertain because of its corruptible and changeable nature and could not yield "true" or "certain" information. In contrast, the supernatural world was immaterial, simple, indivisible, incorruptible, and unchangeable. Instead of being

108. Dee, MP, sigs. ai^v, aiii^r,*i^r ‑ v; discussed in Debus's "Introduction," Dee, MP, p. 10.

109. Dee, MP, sig. *j.   110. Ibid., p. [1^v].

comprehended by the exterior senses through observation, the supernatural world could only be perceived by the mind. Literally above nature, questions about the supernatural levels of the cosmos formulated by the mind could not involve conjecture and speculation, therefore "chief demonstratio[n] and most sure science" could be deduced from them.[111]

It was the supernatural that provided the natural philosopher with certainty and truth and perhaps, by extension, could help him to understand the workings of the natural, corrupt world. This absolute distinction would seem to rule out the value of observation and experimental science in the practice of natural philosophy. Mathematics, however, made observation more reliable, as Dee reminded his readers, because it participated in an intermediary level of the cosmic system. Located between the natural and the supernatural, mathematics occupied the middle ground between two cosmic extremes. Though the mathematical level of the cosmos included "thinges immateriall," they were still "by materiall things hable somewhat to be signified."[112] While mathematical forms were incorruptible and unchangeable, their images or signs were divisible and able to be manipulated and superficially altered. Dee marveled at the "strange participatio[n]" mathematics achieved "betwene things supernaturall, imortall, intellectuall, simple and indivisible: and thynges naturall, mortall, sensible, compounded, and divisible."[113]

The mathematical philosophy of Proclus (fifth c. A.D.) also informed Dee's arguments in the "Mathematical Preface," as can be seen in his annotated copy of Proclus's commentary on the first book of Euclid's *Elements*.[114] Proclus privileged mathematics in much the same way as Dee, and for many of the same reasons. Specifically, Proclus revered the numerical sciences because they could refer to the intangible and incorruptible as well as the sensible. Mathematics was more reliable, therefore, than any branch of natural philosophy that was based on perception.[115] Proclus's philosophy differed from Plato's in the role

---

111. Ibid.   112. Ibid.   113. Dee, MP, n.p.

114. Proclus, *Primum Euclidis Elementorum librum Commentariorum ad universam mathematicam disciplinam principium eruditionis tradentium libri iiii* (Padua, 1560), Roberts and Watson, #266, pp. 1–30. Dee's annotations in Proclus's commentary appear primarily in the metaphysical and philosophical sections of the work. He made no comments on the passages where Proclus discusses the utility of mathematics and numbers or the qualities necessary in the mathematician. See also Clulee, pp. 157–159.

115. Dee's marks of emphasis in his copy of Proclus, p. 6: "nam id quidem ultra intelligentiam obtinent, ut quod evolutum est, & progrediendi vim habet conte[m]plentur: ea vero, quae in ipsis reperitur rationum stabilitate, qu[a]e etiam confutari non potest, opinionem superant. & quod

played by imagination in mastering the essential truths of mathematical philosophy. Proclus emphasized the imagination's intermediary position within the hierarchy of knowing, just as mathematics occupied an intermediary position between the tangible and intangible, of which Dee took note.[116] In these annotations Dee demonstrated his fascination with the concept of mathematical mediation between the sensible and the intelligible, as well as expressing an interest in the ways in which imagination could function as a mathematically relevant feature of the intellect.

Dee resorted not only to pagan and late antique philosophy for the ideas in his "Mathematical Preface," however. He also studied the work of early Christian authors, many of whom were themselves influenced by Neoplatonic philosophy, to support the intermediary place of mathematics. Dee drew on Boethius (d. 524) as an authoritative source on the essentially mathematical nature of the Creation because he accorded mathematics a central role in the most arcane mystery of all: the precise moment of divine creation. Mathematics, Boethius believed, *ordered* the Creation. Dee agreed, as is clear when he quoted from Boethius: "All thinges (which from the very first originall being of thinges, have bene framed and made) do appeare to be formed by the reason of Numbers. For this was the principall example or patterne in the minde of the Creator."[117] Boethius's writings gave Dee a way to reconcile contradictory passages in Genesis and Wisdom, for God thought of the Creation in mathematical terms and instilled that order into every level of the cosmic system through the spoken word.

Boethius's emphasis on pattern ensured that geometry was not altogether forgotten in Dee's "Mathematical Preface." Specifically, Dee be-

---

quidem *ex suppositione* ortum traha[n]t, id fortit[a]e sunt, iuxta prim[a]e scienti[a]e diminutione[m]: quod vero in iis formis constit[a]e sint, *que sine materia* existu[n]t, iuxta perfectiorem sensilium cognitionem."

116. Morrow, "Introduction," in Proclus, p. xxxv, and Dee's emphasis marks in his annotated copy of Proclus, *Primum Euclidis Elementorum*, p. 30: "*At phantasia medium* inter cognitiones obtinens centrum, excitatur quidem a sese, promitque ide, quod sub cognitionem cadit: eo autem q[ue] extra corpus non est, ab illa *vitae* impartibilitae ad partitionem, & intervallum, & figuram, ea, quae sub ipsius cadunt cognitione[m] deducit. Et ideo quicquid noverit, *impressio quedam est, & forma intelligenti[a]e.* Circulum q[ue] una cum suo cognoscit intervallo, ab externa quide[m] materia immunem, *intellectilem vero, quae in ipsa est materiam habentem.* Atq[ue] idcirco *non unus* tantum *in ipsa est circulus*, quemadmodum neq[ue] in sensilibus. Simul nanq[ue] apparet distantia, maius q[ue], & minus, necnon circulorum, ac triangulorum multitudo."

117. Dee, MP, sig. *j.

lieved that geometry was divine because it provided the devout natural philosopher with knowledge of the everlasting forms which had been in the mind of God at the Creation. Geometry, therefore, could "prepare the Thought [of the natural philosopher], to the Philosophicall love of wisdome: that we may turne or convert, toward heavenly thinges [the attention] which now . . . we cast down on base or inferior things."[118] The divine nature of geometry encouraged Dee to give up the ancient label of "geometry" for "megethology," which he believed more accurately captured the theological, metaphysical, and terrestrial uses for the science. "Megethologicall" contemplations, Dee thought, would permit the natural philosopher "to conceive, discourse, and conclude of things Intellectuall, Spirituall, aeternall, and such as concerne our Blisse everlasting: which, otherwise (without Special privilege of Illumination, or Revelation fro[m] heaven) no mortall mans wyt (naturally) is hable to reach unto, or Compasse."[119]

Marvelous revelations were in store for the natural philosopher who devoted himself to the use of intermediary agencies such as mathematics and the imagination. In Dee's "Mathematical Preface" numbers became the specific focus of the natural philosopher's contemplative life and the meditative tool by which natural philosophers

> may both winde and draw our selves into the inward and deepe search and vew, of all creatures distinct vertues, natures, properties, and *Formes*: And also, farder, arise, clime, ascend, and mount up (with Speculative winges) in spirit, to behold in the Glas of Creation, the *Forme* of *Formes*, the *Exemplar Number* of all things *Numerable*: both visible and invisible: mortall and immortall, Corporall and Spirituall.[120]

This passage reflects Dee's ecstatic level of engagement with the essential, numerical components of the cosmos and his hope that the quest for wisdom could be fulfilled through the study of the mathematical sciences, as Plato had suggested when he wrote that wisdom could be attained by "good skill of *Numbers*." Dee elevated the science of arithmetic to a place just below theology in the hierarchy of knowledge because of its profound divinity and purity.[121] "O comfortable allurement, O ravishing perswasion," Dee enthused, "to deale with a Science, whose Subject is so Auncient, so pure, so excellent, so surmounting all creatures, so used of the Almighty and incomprehensible wisdom of the Creator, in the distinct creation of all creatures."[122]

Nevertheless, in the "Mathematical Preface" there is also a glimmer

118. Ibid., sig. aij^v.   119. Ibid., sig. aiij.   120. Ibid., sig. *j–*j^v.
121. Ibid., sig. aj^v.   122. Ibid., sig. *j.

of discontent, a sense that perhaps mathematics alone will not provide the universal science that Dee sought: the first mention of "archemastrie." Dee described archemastrie as a universal science that was experimental, revelatory, earthly, and divine.[123] Also, he stated his belief that the "chief Science, of the Archemaster, (in this world) as yet known, is an other (as it were) OPTICAL Science" – quite possibly the use of a stone to see angels.[124] In earlier common usage, the word *archemastry* derived from a corruption of the word *alchemy*, but in the sixteenth century the word came to be used for high skill or mastery of the applied natural sciences or mathematics. Dee considered his "Archemastrie" to be the unification of all branches of natural philosophy into a single discipline that explored the propagation of rays, employed mathematical aspects of optics, depended upon astrology to capture astral radiation, and utilized a highly refined stone resulting from an alchemical experiment. When used together, these natural philosophical skills and techniques would raise the natural philosopher from the contemplation of transitory terrestrial conditions to a state of communication with celestial truths.

Dee's "Archemastrie," the experimental science related to spiritual experience, divination, and natural philosophy, may indicate that he was already attempting to communicate with the angels.[125] As early as 1570, based on the evidence from the "Mathematical Preface," Dee's intellect was engaged with the possibilities of a new optical, experimentally based natural philosophy which he would later practice in his angel conversations. Steeped in the optics of his predecessors Bacon and Grosseteste, Dee formulated a means of contacting angels that did not rely on conjurations or binding spells, but on simple prayer and the use of a crystal stone to collect the occult rays which all major theological and philosophical authorities agreed were vehicles for angelic, as well as astral, influences. Dee's use of the crystal stone was more complex than we might initially perceive, due to both his expertise in optics and his inter-

---

123. Nicholas H. Clulee, "At the Crossroads of Magic and Science: John Dee's Archemastrie," in *Occult and Scientific Mentalities in the Renaissance*, ed. Brian Vickers (New York: Cambridge University Press, 1984), pp. 57–71.

124. Dee, MP, sig. Aiii[v].

125. See Nicholas H. Clulee, "At the Crossroads of Magic and Science: John Dee's archemastrie," *Occult and Scientific Mentalities in the Renaissance*, ed. Brian Vickers (New York: Cambridge University Press, 1984); pp. 57–71. Calder also makes this association between optics and Dee's mysterious "archemastrie," see I. R. F. Calder, "John Dee Studied as an English Neoplatonist" (diss., The Warburg Institute, London University, 1956), 1:781.

est in the metaphysics of light. The revelatory power of geometry in the "Mathematical Preface" echoes the revelatory power of the *monas* in the *Monas hieroglyphica* and the revelatory power of the celestial rays to awaken understanding and cause movement toward the divine in the *Propaedeumata aphoristica*. The "Mathematical Preface" also contains a prophetic glimmer of Dee's escalating interest in angelic communication during the 1570s. Angels were, after all, akin to mathematics in their intermediary nature and function. In grasping the valuable step toward divinity that mathematics provided, Dee became increasingly committed to all intermediary cosmic agencies – the mathematical, linguistic, optical, alchemical, and imaginative. The common point toward which all these interests drew him was the angel conversations. Rather than observing, deciphering, or ordering nature, Dee devoted himself to *understanding* the natural world as fully and completely as God would allow.

3

Climbing Jacob's Ladder

Angelology as Natural Philosophy

Unlike other things alive –
Birds that wing their way through air
Fish that in chill waters dive
Beasts that tread the earth
and leave their footprints there –
Man, by reason made aware,
Solely judges and discerns;
Man approaches angels by the things he learns.
> —Francesco Stelluti
> in Galileo Galileo's *Assayer*, 1623

For a decade after the "Mathematical Preface" a veil falls over our knowledge of John Dee's interest in intermediary agencies, his commitment to ascending to the supernatural levels of the cosmos, his fascination with universal sciences, and his provocative "archemastrie." The veil lifts in 1581, just as Dee was about to enter into his collaboration with Edward Kelly, when he addressed a prayer to God explaining how and why he communicated with angels. Today it stands as the preface to the first surviving angel diary.[1] The preface captures another important moment in the history of the conversations, in this case not an ending but a beginning, a moment of reflection for Dee as he made a transition between two scryers and renewed his attempts to climb to the heavens and achieve certain wisdom. The political difficulties Dee would later experience in the Rudolphine court in Prague, the serious danger posed by his overexposure to so many curious eyes, and the struggles he would have with his irascible new scryer were all in the future. Dee's studies and publications supplied the materials for his ladder to the heavens and informed his outlook, but they were no longer sufficient. In 1581 Dee's long and painful intellectual crisis was drawing to a close,

---

1. Dee, *AWS* II: 8–12.

and its residue remains in the account he made of his intellectual despair and frustration.

The roots of Dee's intellectual crisis extend back to his early career as a scholar when he had studied "in many places, far & nere, in many bokes, & sundry languagis [*sic*]." Dee described how he had conferred "with sundry men" and had with his "owne reasonable discourse labored." This traditional course of study had failed him; it had not brought him the mastery of the Book of Nature he desired. He labored on, collecting his library and publishing his own natural philosophy, but at last he concluded he "could fynde no other way, to . . . true wisdome" without the "extraordinary gift" of divine revelation. Though Dee confessed that he had been making "special supplications" to the angel Michael from 1569, these requests had failed too.[2] Dee fell deeper into despair, and might even have contemplated suicide, as is suggested in his resolution "to leave this world presently . . . so, I might in spirit enioye the bottomles fowntayne of all wisdome."[3] Something – perhaps contact with a scryer – saved him from a final desperate act, and he turned once more to his library for inspiration on how to circumvent his problems and achieve wisdom.

Dee was not looking for wisdom itself on his library shelves: he was looking for a new *method* to acquire it.[4] In 1581, as he began to work with Kelly, Dee was reminded of passages in the Bible that told how God sent his "good angels" to Enoch, Moses, Jacob, Esdras, Daniel, and Tobit, "to instruct them, informe them, help them . . . in worldly and domesticall affaires . . . and sometimes to satisfy theyr desyres, dowtes & questions of thy Secrets." Armed with these examples from the Book of Scripture, Dee drew on religious texts for a method to facilitate communication with God and His angels. The Bible and its theological commentaries prompted him to use a "Shewstone, which the high priests

2. A firm date for the inception of the angel conversations has yet to be established. While Dee records in Sloane MS 3188 that he began to pray to the angel Michael in 1569, Nicholas Clulee discovered in Psellus's "De Daemonibus," a place where Dee had noted the date "1567" in the margins alongside a passage on divination. This suggests that Dee made some attempt to contact spirits prior to 1569. See Clulee, p. 141, and Dee's notes in Ficino's *Index eorum*, p. 52ᵛ, Roberts and Watson, #256.

3. Dee, *AWS* II: 341.

4. For a discussion of this distinction between the recovery of ancient wisdom and the recovery of ancient methods to achieve wisdom, see Peter Dear, *Discipline and Experience* (Chicago: University of Chicago Press, 1995), especially pp. 118–122.

did use ... wherein they had lights and Judgements in their great dowtes." Because of its biblical antecedents, Dee's showstone allowed him to draw a sharp distinction between his conversations with angels and contemporary magical practices. With the Book of Scripture to support him, Dee could say sincerely that he had "allwayes a great regarde & care to beware of the filthy abuse of such as willingly and wetingly, did invocate and consult diverse sorts [of] Spirituall creatures of the damned sort." Instead, Dee used "harty prayer" and the skills of scryers like Edward Kelly who were capable of seeing "Spirituall apparitions, in Christalline receptacles, or in open ayre" which harked back to his optics and "archemastrie."[5]

Finally, Dee discussed the reasons he contacted the angels. First, he wanted to achieve "pure and sownd wisdome and understanding of ^some of^ thy [God's] truths naturall and artificiall." Dee's emphasis on "truths naturall and artificiall" is important because it links the contents of his angel conversations to natural philosophy and expresses his confidence that the angels could further those interests. Subsequent remarks indicate that there was yet a deeper purpose underlying the angel conversations: he wanted to communicate with angels so that God's own "wisdome, goodness and powre bestowed in the frame of the world might be browght, in some bowntifull measure under the Talent of my Capacitie." The acquisition of wisdom was not solely for Dee's benefit, but for God's "honor & glory, & the benefit of thy Servants, my brethern and Sistern." Here Dee reveals his belief that an angelically disclosed natural philosophy was not only possible, but desirable because it might restore humanity's once-privileged place in the natural world, harness the inherent powers of the Book of Nature for human benefit, and honor God.

These were lofty aspirations, and it is not surprising that Dee despaired of natural philosophy, given his grand expectations. As this chapter will explain, Dee believed that a reformation of the human sciences and human knowledge was needed to bridge the gap between his expectations and the decayed state of the Book of Nature and natural philosophy. The angel conversations not only converged with broader late-sixteenth-century cultural contexts such as humanism, they also correlated with a more specific intellectual context provided by the per-

5. For the broader cultural and intellectual context of Dee's angel magic, as well as a finer sense of why Dee made these distinctions, see D. P. Walker, *Spiritual and Demonic Magic: From Ficino to Campanella* (Notre Dame, IN: University of Notre Dame Press, 1975), passim, where the varieties of spiritual and demonic practices are outlined.

ceived inadequacies of natural philosophy during the period. Dee lived during the nascent years of the period now known as the "scientific revolution," when other sixteenth-century voices were also calling out for philosophical and methodological reform.[6] Some natural philosophers clamored for order among the branches of natural philosophy – an impulse that culminated in the disciplinary boundaries of modern science. Others searched for the "pure" doctrines of natural philosophy possessed by the ancients, stripped of the corruptions, inaccuracies, and deterioration on which humanists focused their attentions. Others, like Dee, attempted to craft "universal sciences" that would combine ideas of doctrinal order and purity espoused by other reformers with changes in method and approach. Because Dee had not been able to make the climb upward using the natural philosophy available to him, he established a new, universal science based on optically facilitated conversations with the perfect, celestial world. Dee's "ladder" to the heavens – the angel conversations – enabled communication between separate parts of the cosmic system and offered the hope of intellectual and spiritual perfection to an increasingly frustrated and anxious world.

John Dee was not the first, nor the only, scholar of the period to aspire to certain knowledge of nature and experience frustration when he failed to achieve it. Henry Cornelius Agrippa underwent a similar crisis of intellectual confidence earlier in the century. Agrippa's *De incertitudine et vanitate scientiarum* discussed the inability of philosophers to reach wisdom through study of the existing arts and sciences.[7] His critics pointed out that Agrippa's book disowned the very activities he had championed earlier in his career: humanism, scholarship, and critical inquiry. One of the most intriguing features of Agrippa's diatribe comes after a scathing review of all fields of human knowledge and applied reason, when he concludes that only "prophetical" divinity (revelation) and "interpretive" divinity (a study of God's scriptures), coupled with a

6. See Dear, *Discipline and Experience*, passim; Mark Greengrass, Michael Leslie, and Timothy Raylor, eds., *Samuel Hartlib and Universal Reformation: Studies in Intellectual Communication* (Cambridge: Cambridge University Press, 1994), passim.

7. Several English translations of this work were completed in the sixteenth and seventeenth centuries. All quotations here are from Henry Cornelius Agrippa, *Of the vanitie and uncertaintie of the Artes and Sciences*, trans. James Sanford (London, 1569). Calder noted a similarity in mind-set between Dee prior to his engagement with the angel conversations and the Agrippan crisis. See Calder, "John Dee as English Neoplatonist," I:777. There is no record that Dee owned Agrippa's *De incertitudine*, though many other items from the Agrippan corpus appear in his library catalogue.

life devoted to God, can bring a philosopher certain knowledge and wisdom. Agrippa exhorted his readers to "descend into your selves you whiche are desirous of the truthe, departe from the clouds of mans traditions, and cleave to the true light: beholde a voice from Heaven, a voice that teacheth from above."[8]

Agrippa might have inspired Dee to *build* Jacob's ladder, and also to *climb* it and attend to voices teaching "from above," but he was not the only philosopher to help Dee conceptualize his angel conversations. A closer examination of the books Dee purchased and annotated between 1544 and 1581 reveals that an array of ancient and medieval ideas informed his specific strategy for communicating with angels. Ancient and medieval natural philosophers Dee held in high esteem, such as Roger Bacon and Raymond Lull, had already suggested that a universal science could provide the greatest insights into the problems of natural philosophy. Dee found that the most appealing option facing him was to resolve the intellectual difficulties posed by two competing, unstable, and unreadable "texts" – the Books of Scripture and Nature – through the medium of a completely new universal science. Whereas Bacon posited that a truly universal science must be based on optics and the behavior of light, and Lull felt a system of numbers and letters could clarify and organize the cosmos, Dee looked directly to the incorruptible celestial spheres for the strong foundation of his final attempt to craft a universal science – the angel conversations. Unlike his earlier attempts in the *Propaedeumata aphoristica*, the *Monas hieroglyphica*, and the "Mathematical Preface," Dee hoped the angel conversations would be capable of answering all questions posed by the Book of Nature, and would be authoritative in other branches of the arts and sciences as well.

Dee's Angelic Cosmology

Dee's decision to look to the angels and celestial spheres for answers to his natural philosophical questions is hardly surprising. A quick glance at the art, literature, and architecture of the early modern period reveals that angels were everywhere. They decorated the stained glass windows, interior beams, and exterior arches of great cathedrals and humble village churches while music-making angels adorned altars and choirstalls.[9]

8. Agrippa, *On the Vanitie*, p. 187ᵛ.
9. For ecclesiastical and secular art, see Gunnar Berefelt, *A Study on the Winged Angel. The Origin of a Motif*, trans. Patrick Hort (Stockholm: Almquist and Wiksell, 1968); Nancy Grubb, *Angels in Art* (New York: Artabras, 1995); Valentine Long, *The Angels in Religion and Art* (Paterson,

Secular architects filled domestic spaces with ceilings and wall panels depicting the zodiac, classical gods and goddesses, and armies of winged creatures from majestic avenging angels to adorable putti.[10] Angels appeared in the popular dramas based on Biblical stories known as mystery plays that were produced throughout Europe during the sixteenth century.[11] As God's cosmic helpers, angels appeared in tapestries on the walls of great houses and in the frontispieces of books, turning the great mechanical crank that kept the celestial spheres spinning in perpetuity.[12]

Dee's belief in the potential value of angelic guidance was shaped by his understanding of the structure and workings of the cosmos, which allowed for both natural laws and a providential deity, for both predictable and miraculous events. In his cosmology, angels were aspects of faith as well as significant intermediary features of the natural and supernatural worlds. What distinguished Dee's angel diaries from other early modern records of experiments involving angels was his familiarity with the traditional literature on angels, his fascination with the power of intermediaries, and his deeply held belief that angels were divine messengers.[13]

The sixteenth-century cosmos was hierarchical and stretched in a chain of influences from God through the angels and celestial spheres to the sublunar world of human beings, animals, and plants. Dee and his contemporaries were interested in angels because, in their cosmology, the most important relationships were those between God, the angels, and human beings. Human beings, hampered by Adam's original sin, were unable to communicate with God and His creatures and had to

NJ: St. Anthony Guild Press, 1970); Peter Lamborn Wilson, *Angels* (London: Thames and Hudson, 1980).

10. Prominent among early modern examples are the frescoes commissioned by the Medici. See Janet Cox-Rearick, *Dynasty and Destiny in Medici Art: Pontormo, Leo X, and the Two Cosimos* (Princeton, NJ: Princeton University Press, 1984).

11. See Deborah E. Harkness, "Shows in the Showstone: A Theater of Alchemy and Apocalypse in the Angel Conversations of John Dee," *Renaissance Quarterly* 49 (1996): 707–737, pp. 722–724.

12. For vivid illustrations of this idea, see S. K. Heninger, Jr., *The Cosmographical Glass: Renaissance Diagrams of the Universe* (San Marino, CA: The Huntington Library, 1977), passim.

13. To compare Dee's angel conversations with contemporary attempts to contact angels, see the following manuscripts. At the British Library, Harleian MS 2267, f.61; Additional MS 36674, ff. 58–63 and 167–196; Sloane MS 3702; Sloane MS 3934, ff. 1–135. At the Bodleian Library, MS Rawlinson D253; MS Add. B.1; MS Ballard 66, ff. 27, 29, 31. At Cambridge University Library, MS Add. 3544; MS Li.1.12, f. 21v.

struggle to emulate the angels, who were perfect and unaffected by original sin and the moral fall. In the world promised after the Apocalypse, angels and humanity would live in harmony with nature, in perfect communion with God and all levels of the cosmos.[14]

Dee's ideas derived not only from general cultural beliefs of the sixteenth century but also from well-defined intellectual traditions. Books concerning angels made up a significant part of his library, and as early as 1557 he had several texts which supported his later interest in communicating with angels. These included a collection of Neoplatonic and Hermetic texts edited by Ficino, the works of Dionysius the Areopagite, and four essential works on magic and the Christian cabala by Francesco Giorgi, Henry Cornelius Agrippa, Johannes Reuchlin, and Johann Trithemius. Although the latter four have not been recovered, Dee extensively annotated the texts edited by Ficino and the works of Dionysius. They provide helpful clues regarding particular aspects of angelology that interested him.

Dee's marginalia also suggest that he not only *read* most of the works on angels available in the sixteenth century, he *mastered* them.[15] As a result, Dee seldom spent much time trying to recall which author discussed an idea relevant to his angel conversations, but was able to think of supporting texts quickly and, for the most part, accurately. He often noted the similarities between information that the angels revealed in the conversations and the work of early modern occultists like Trithemius,

14. Authors who have emphasized the importance of early modern cosmological beliefs include E. M. W. Tillyard, *The Elizabethan World Picture* (New York: Penguin, 1984); Arthur O. Lovejoy, *The Great Chain of Being: A Study of the History of an Idea* (New York: Harper and Row, 1960); C. S. Lewis, *The Discarded Image: An Introduction to Medieval and Renaissance Literature* (Cambridge: Cambridge University Press, 1967); and S. K. Heninger, Jr., *Touches of Sweet Harmony. Pythagorean Cosmology and Renaissance Poetics* (San Marino, CA: The Huntington Library, 1974); Fernand Hallyn, *The Poetic Structure of the World: Copernicus and Kepler* (1990), trans. Donald M. Leslie (New York: Zone Books, 1993); James Miller, *The Cosmic Dance* (Toronto: University of Toronto Press, 1986); N. Max Wildiers, *The Theologian and His Universe: Theology and Cosmology from the Middle Ages to the Present*, trans. Paul Dunphy (New York: Seabury Press, 1982); René Roques, *L'Univers dionysien: Structure hiérarchique du monde selon le Pseudo-Denys* (Paris: CERF, 1983).

15. His mastery might have encouraged Dee to write a book on the nature of angels and spirits, now lost, as Dee noted in his personal diary on 7 July 1600 that, "as I lay in my bed, it cam into my fantasy to write a boke, 'De differentiis quibusdam corporum et spirituum.' " Dee, *PD*, p. 62.

Agrippa, Reuchlin, and Peter d'Abano, displaying an encyclopedic command of the relevant literature.[16] In addition, Dee kept works he frequently referred to in the study where the angel conversations took place, rather than in his more public library.[17] When he left England for the Continent, he took a large portion of his great collection with him to central Europe. Dee marked for travel so many of the books that discussed angels – like Trithemius's *De septem secundeis* – we can only conclude that he found them helpful in his angel conversations.[18]

Dee also drew on less esoteric printed sources for his angelology, including the Bible.[19] In the Bible, angels appear sparsely in the Old and New Testaments, and more regularly in the Apocrypha and Pseudepigrapha, which were still a part of printed Bibles during Dee's lifetime. Important to Dee's establishment of the angel conversations as a universal science was the long and authoritative biblical history of the role angels played in helping humans to understand God's plan through visionary revelations. From his Bible, Dee would have been familiar with angels as divine messengers, guides on revelatory journeys, and agents of the Apocalypse.[20]

Biblical accounts provided later Christian commentators with ample

16. See Dee, *AWS* I:30–34 (where both Agrippa and Reuchlin were mentioned), 36–39 (where Peter d'Abano was mentioned); *TFR*, pp. 12–13 (where Trithemius was mentioned); Clulee, p. 211.

17. Dee, *AWS* II:39.

18. For a discussion of the "T" designation in the Trinity College copy of Dee's library catalogue, see Roberts and Watson, p. 49.

19. William George Heidt, *Angelology of the Old Testament* (Washington, DC: Catholic University of America Press, 1949); Charles Fontinoy, "Les Anges et les démons de l'Ancien Testament," in *Anges et démons*, ed. Julien Ries and Henri Limet, Homo Religiosus 14 (Louvain-le-Neuve: Centre d'Histoire des Religions, 1989), pp. 117–134; Joseph Ponthot, "L'Angélologie dans l'Apocalypse johannique," ibid., pp. 301–312.

20. Although one of the most prominent examples of this tradition is the story of Jacob's ladder, there are also important references to angelic messengers in the New Testament. In Matthew 1:20, an angel allays Joseph's fears about Mary's pregnancy, and in Luke 1:26, the angel Gabriel foretells the birth of Christ to Mary. In the Apocrypha, angels typically function as guides on revelatory journeys, as in the case of Tobit (when the angel Raphael appears to Tobit to serve as his son's spiritual guide) and 2 Esdras (when the angel Uriel shows Esdras secrets about the created world). Throughout Revelations, angels play a more active role: announcing the end of days, bringing souls to judgment, and sealing Satan in the Abyss for a thousand years.

material for discussion, and Dee relied heavily on these works for their synthetic treatment of angelic natures. Some of the sources Dee used to inform his angelolgy were philosophical, some theological; and still others were concerned with specific properties of the Book of Nature.[21] His most important source was undoubtedly Dionysius the Areopagite, on whom so many other Christian commentators had depended.[22] Dee owned the entire Dionysian corpus, including a copy of his complete works with a commentary by Jacques Lefevre DÉtaples (Venice, 1556), and an additional copy of *De mystica theologica* with a commentary by the German mystic Eck (Rome, 1525).[23] The copy of *De mystica theologica* is intriguing because it demonstrates that Dee was as interested in contemporary commentaries on the work of Dionysius as he was in the text itself.[24] But some contemporary commentaries were missing from Dee's collection, most notably Ficino's much lauded edition of the Dionysian corpus, which had been instrumental in the reintroduction of Platonic ideas into western Europe in the fifteenth century. Instead, the editions Dee owned contained more current, post-Reformation commentaries that blended Dionysius's ancient ideas with the most up-to-date humanistic scholarship and theology.[25]

21. I. R. F. Calder includes a detailed exposition of possible sources for Dee's angelology in his two-volume thesis, which focuses on Platonic and Neoplatonic influences. Calder, "John Dee Studied as an English Neoplatonist," 1:739–833.

22. For Dionysius's influence on cosmology and theology, see René Roques, *L'univers dionysien*, and D. Knowles, "The Influence of Pseudo-Dionysius on Western Mysticism," in *Christian Spirituality: Essays in Honour of Gordon Rupp*, ed. P. Brooks (London: SCM Press, 1975), pp. 79–94, passim. Throughout the Middle Ages scholars and theologians believed that Dionysius was the disciple of Saint Paul who was responsible for converting the Athenians to Christianity. Dionysius's works were redated by Lorenzo Valla in the fifteenth century, making this an impossibility. However, this redating had little effect on scholarship for over a century. On Dionysius's association with Saint Paul, see Guntriem G. Bischoff, "Dionysius the Pseudo-Areopagite, the Gnostic Myth," in *The Spirituality of Western Christendom*, comp. E. Rozanne Elder (Kalamazoo, MI: Cistercian Publications, 1976).

23. Roberts and Watson, #975 and #271, respectively.

24. For an analysis of this commentary, and its associations with the theology of Nicholas of Cusa, see Georgette Epiney-Burgard, "Jean Eck et le commentaire de la *Théologie Mystique* du Pseudo-Denys," *Bibliothèque d'humanisme et renaissance* 34 (1972): 7–29.

25. See Karlfried Froelich, "Pseudo-Dionysius and the Reformation of the Sixteenth Century," in *Pseudo-Dionysius: The Complete Works*, trans. Colm

The Dionysian corpus was second only to the Bible and the works of Boethius in the number of translations, editions, and further commentaries it generated throughout the Middle Ages and the early modern period.[26] It was Dionysius, for example, who first elaborated the three hierarchical divisions of angels, each containing three separate orders, or choirs, of angels. The highest order was comprised of Seraphim, Cherubim, and Thrones, who surrounded God and sang his praises. On the next level Authorities, Dominions, and Powers directed their activities toward God, standing "with their faces to Him and their backs to us."[27] Finally, the angels who ministered to humanity were in the lowest order: the Principalities who guarded over nations, and the Archangels and Angels who delivered messages to humanity. The idea that angels (at least angels of the lowest level of the celestial hierarchy) were specifically created to communicate with human beings had a particularly strong appeal to subsequent scholars – so much so that Martin Luther, an opponent of Dionysius's Platonism and Neoplatonic mystical piety, encouraged his followers to "shun like the plague that 'Mystical Theology' of Dionysius and similar books" which "taught that humans can converse and deal with the inscrutable, eternal majesty of God in this mortal, corrupt flesh without mediation."[28]

The possibility of direct communication with God was contested during the stormy days of the Reformation. Because angels were linked to human redemption and salvation and thus to the elect, references to angels appeared in Protestant as well as Catholic treatises. The importance of angels was magnified by their numerous appearances in the apocalyptic Revelation of John the Divine. Angels were expected to fulfill many roles in the final days, including communicating to the chosen, in advance, the full scope of the events. More important, however, the angels provided a model for exemplary human conduct. Stripped of earthly needs for food and sustenance and free from sinful human desires and impulses, the perfect, orderly lives of the angels provided a stable anchor in the confusion of the Reformation period. People interested in the ways angels could serve as models for human conduct needed to look no further than one of the many editions of the Dionysian corpus for guidance. In Dionysius's scheme, angels helped to redeem humanity "by prescribing roles of conduct, by turning them from wandering and

Luibheid, foreword, notes, and translation collaboration by Paul Rorem; preface by René Roques; introductions by Jaroslav Pelikan, Jean LeClercq, and Karlfried Froelich (London: SPCK, 1987) pp. 37–38.

26. Ibid., p. 33.   27. Lewis, *The Discarded Image*, p. 72.
28. Quoted in Froelich, "Pseudo-Dionysius," p. 44.

sin to the right way of truth, or by coming to announce and explain sacred orders, hidden visions, or transcendent mysteries, or divine prophecies."[29]

The angels' function as role model was an important feature of six-teenth-century theological and cosmological beliefs, but the angels had also held positions of wider cosmological importance. Many of these roles had been expansively treated in the ancient, Arabic, and Jewish traditions – which only made them more problematic to orthodox Prot-estants and Catholics. When pagan, Jewish, and Arabic angels joined with Christian angels, they became omnipresent and necessary features of the cosmos on a structural-functional level as well as agents in human salvation. No longer limited to the functions of messengers or redemp-tive role models for humans to follow, angels became Christianized planetary intelligences responsible for moving the celestial spheres. Though some writers continued to try to separate angels and planetary intelligences, most found themselves unable to keep the two categories distinct.[30] Even Dee, who often took issue with inappropriate terminol-ogy, noted in his copy of Azalus's encyclopedia that planetary angels or "intelligences" moved the heavenly bodies.[31]

Once scholars had linked angels and celestial intelligences, further connections were made between angels and the Platonic "forms" some philosophers believed were embedded in the Book of Nature. The asso-ciation of angels and forms, in turn, contributed to early modern occult systems of sympathy and antipathy that drew heaven and earth to-gether.[32] By Dee's lifetime, angels played a key role in the maintenance of cosmic order and were credited with creating the music of the spheres, grasping the divine language of creation and understanding the divine will. Finally, angels were materially related to the natural world and humanity through the soul and essential forms. By the sixteenth century, angels provided a vitalistic, cohesive principle for the cosmos similar to early ideas about gravity – anything that happened in the cosmos that

29. Pseudo-Dionysius, "The Celestial Hierarchy," p. 157.
30. Richard C. Dales, "The De-Animation of the Heavens in the Middle Ages," *Journal of the History of Ideas* 41 (1980): 548; Edward Grant, *Planets Stars, and Orbs: The Medieval Cosmos, 1200–1687* (1994) (Cambridge: Cambridge University Press, 1996), pp. 526–545.
31. Dee's emphasis in Azalus, p. 17ʳ. Dee noted in the margins "Intellige[n]tiae orbes caelestes movent" and underlined passages regarding this angelic function.
32. Harold P. Nebelsick, *The Renaissance, the Reformation and the Rise of Science* (Edinburgh: T. & T. Clark, 1992), p. 135.

was inexplicable through mechanical or physical means was attributed to an angelic agency working at God's behest.

At the time Dee was engaged in his angel conversations this eclectic concept of an angel was firmly rooted in works of theology and natural philosophy. Angelology was particularly relevant to those who were attracted to the occult aspects of the Book of Nature, or their magical manipulation. This was the great danger of the syncretistic conception of angels, with its pagan and heretical roots: it could be construed as demonic. Given the range of authors who contributed to the hybrid, early modern Christian angelology, disagreement over this issue was prevalent – as were charges of sorcery, conjuring, and enchantment against those interested in communicating with angels. Natural philosophers interested in the cosmos's occult properties fueled the controversy surrounding the potentially demonic connotations of angelic communication by emphasizing the enormous possibility for knowledge and wisdom that existed precisely because of the angels' liminal, intermediary position between the created and spiritual worlds. One of Dee's contemporaries, the natural philosopher Francesco Giorgi, reminded his readers, in *De harmonia mundi* (1545) that, just as God descended to the world by emanation through the angels, so humans might use the angels as a ladder to ascend to God.[33] Such conclusions were bound to cause concern in a culture both fascinated and repelled by the idea of contacting spiritual beings. Giorgi was not alone in his conviction that angelic intermediaries could resolve many mysteries in the Book of Nature. Trithemius's *De septem secundeis* contained a similar mixture of astrology, cabala, and angelology.

Dee was interested in the redemptive possibilities of communication with angels, despite its dangers. The marginalia in his copy of Dionysius's works indicate that Dee adopted the celestial order or hierarchy filled with angels and was concerned with mastering the specific nuances of the system. He noted, for example, that the number of angels within each hierarchy was "innumerable," despite scriptural references to "a thousand times a thousand" or "ten thousand times ten thousand."[34] He

---

33. Quoted in Thorndike, V:451. Dee's copy of Francesco Giorgi's *De harmonia mundi* (Paris, 1545) is Roberts and Watson, #221/B91; it has not been recovered.

34. Dee's emphasis in Dionysius, *Opera* (Venice, 1556), Roberts and Watson, #975, p. 47ᵛ [sig. F6ᵛ]: "At hoc item spiritali [*sic*] discussione & magisterio dignu[m] puto, quamobrem scripturae sanct[a]e traditio, de angelis loque[n]s *millia* millium esse dicat, & *decies millies denamillia, supremos humanos numeros* in seipsos involvent atque multiplicans, perq[ue] hos

emphasized that the celestial hierarchy was more than just an organizational scheme; it also represented different modes of active *participation* in the cosmic system. Some angels participated more fully in the divine plan than other angels, and all angels were more active in the cosmos than was humanity. Still, Dee believed that all members of the hierarchy, both divine and terrestrial, were *"fellow workmen for God and a reflection of the workings of God"*[35] because the hierarchical continuum ensured that all parts of the Book of Nature, high and low, were *"as like as possible to God."*[36] But the angels of the celestial hierarchy were *"images of God in all respects, . . . clear and spotless mirrors* reflecting the glow of primordial light and indeed of God himself" whereas human beings were not capable of such perfection.[37]

Dee's interest in the Dionysian system was significant for his natural philosophy because angels could be linked to the vast web of correspondences that knit the cosmos together. The Dionysian celestial hierarchies could be incorporated into existing lists of sympathetic and antipathetic relationships, and in two places Dee listed correspondences between angels, the twelve signs of the zodiac, the ten spheres, the seven planets, the four elements, and the four winds.[38] These lists demonstrate his familiarity with angels from a variety of traditions, and also his belief that angels were linked to the natural world through occult connections and influences. Though the exact sources for his elaborations are difficult to trace – most early modern authors drew on a variety of authors for their ideas without citation – there are some notable exceptions where he credits an author or refers to a text.

Often, the authors are prominent in what we call the occult sciences. Occult natural philosophers occupied a special place in Dee's library and in the formation of his angelology, providing him with a synthesis of lore about angels drawn from Islamic, Jewish, and Christian authors. In

coelestium substantiarum ordines nobis *esse innumerabiles* apertis rationibus pa[n]dens. . . ."

35. Dee's emphasis in Dionysius, *Opera*, ". . . *dei cooperatore[m] fieri, divinamque in se operationem quantum potest palam cunctis ostendere . . . ,"* sig. B3ʳ. Dionysius, "Celestial Hierarchy," p. 154.

36. Dee's emphasis in Dionysius, *Opera*, ". . . deo *quantum fieri potest similem evadere, unumq[ue] cum illo fieri,"* sig. B2. Ibid.

37. Dee's emphasis in Dionysius, *Opera*, ". . . *eos etiam qui secum divina sectantur, dei signa & imagines efficit, ac perlucida specula & omne labe pura* dignaque quibus principalis illius ac divinae lucis suavissimus radius influat . . . ,"* sig. B2ᵛ. Ibid.

38. Bodleian Library Ashmole MS 1790, f. 30ᵛⁱ see also Dee's annotations in Azalus, *De omnibus rebus naturalibus*, p. 66ʳ.

the angel diaries, Dee stated there were angels who governed particular professions, for example, and credited the doctrine to the occultist Agrippa.[39] Agrippa may not have been the first to think of such a relationship, but this example elucidates the many filters through which Dee's ideas about angels passed as he crafted his rationale for the angel conversations.

Dee found the works of another occult philosopher, Agrippa's teacher Johann Trithemius, deeply compelling, and he owned many of the abbot's works. Trithemius, who became engrossed with the works of Dionysius in 1496, depended on him for his ideas about angels.[40] The work of Trithemius provided Dee with information about the various levels of the cosmic hierarchy as well as ways in which angelic knowledge could be put to practical use by natural philosophers. Dee heavily annotated and underlined Trithemius's *Liber octo quaestionum* (1534), which was concerned with the spiritual life of humanity and the place of angels in the cosmic system.[41] Dee owned another of Trithemius's works, *De septem secundeis* (1545), which contained specific information about the angels who governed the planets and argued for an angelically governed periodization of world history.[42] In addition, Dee acquired a manuscript copy of Trithemius's *Steganographia* when he was on the Continent in 1563, believing that it was "so needfull and comodious, as in humayne knowledg [sic], none can be meeter."[43] The *Steganographia* introduced a method of long-distance communication based on the seven planets and their guardian angels "according to the tradition of the wise men of old."[44] To have utilized the system, the practitioner needed to master technical astrology, especially the exact motions of the planets.[45] Though Dee's interest in Trithemius's system has always been linked to his pos-

---

39. Dee, *AWS* II: 28–34.   40.   Froelich, "Pseudo-Dionysius," p. 35.

41.   Johannes Trithemius, *Liber octo quaestionum, quas illi dissolvendas proposuit Maximilianus Caesar* (Cologne, 1534), Roberts and Watson, #897.

42.   Unfortunately, though Dee's annotations were heavy, the book was rebound, and in the process Dee's marginal notes were damaged; now only a few letters remain in the margins. Johannes Trithemius, *De septem secundeis* (Frankfurt, 1545), Roberts and Watson, #678-T.

43.   See John Dee, "Letter to Sir William Cecil, 16 February 1562/63," ed. R. W. Grey, *Philobiblon Society Bibliographical and Historical Miscellanies* 1 (1854): 1–16.

44.   For accessibility I quote from the modern English translation: Johannes Trithemius, *The Steganographia of Johannes Trithemius*, ed. Adam McLean, trans. Fiona Tait, Christopher Upton, and J. W. H. Walden (Edinburgh: Magnum Opus Hermetic Sourceworks, 1982), p. 97.

45.   Trithemius, *Steganographia*, pp. 104–106.

sible interest in cryptography, there is no evidence to suggest that anyone in Elizabeth's government considered the work anything but a curiosity.[46]

It is clear from Dee's marginalia that he was fascinated by the works of the principal occultists of his time. His angelology was not, however, solely based on the work of occult philosophers. Pompilius Azalus's encyclopedia, *De omnibus rebus naturalibus* (1544), provided another source for Dee's angelology, and though it was not the most original work he consulted, it is indicative of the thumbnail sketches then available. Azalus's work contained, as the title suggested, a discussion of all aspects of the terrestrial and celestial worlds, and was intended to be a useful compendium of information about the cosmos rather than a sophisticated theological or natural philosophical treatise. Nonetheless, Azalus drew on sources from the Christian tradition such as Dionysius, as well as natural philosophical sources.

Dee's copy of *De omnibus rebus naturalibus* is heavily annotated in several sections – most noticeably in the two chapters on angels.[47] He was specifically interested in Azalus's remarks on guardian angels, noting that everyone had a guardian angel who watched over him, inspired him, and stimulated him to do good works.[48] This idea was far from novel, but it could be pressed into greater service by a natural philosopher interested in the role that angels played in the cosmic system. One of Dee's fellow natural philosophers, Girolamo Cardano, who met Dee on a dock in Southwark in late 1552 or early 1553, demonstrates how such an idea could be refined.[49] Cardano believed that guardian angels

46. Jim Reeds explores the cryptological significance of Trithemius's work in "Solved: The Ciphers in Book III of Trithemius's *Steganographia*," forthcoming. Luigi Firpo, using Dee's interest in Trithemius's *Steganographia*, suggests that Dee was an agent for the Crown, in "John Dee, scienziato, negromante e avventuriero," *Renascimento* 3 (1952): 25–84. I concur with French, who argues that any discernable veneer of cryptography was intended to hide the "cabalist-angel magic" that was the true focus of the work (*John Dee*, pp. 36–37).

47. Dee's copy of the work is Roberts and Watson, #134. Azalus's work was included in Dee's continental traveling library.

48. Dee's annotations in Azalus, p. 18ʳ.

49. Girolamo Cardano, *The Book of My Life (De vita propria liber)*, trans. Jean Stoner (London: J. M. Deut & Sons, 1931), pp. 240–247. See Roberts and Watson, p. 75, for the meeting between Dee and Cardano, and Roberts and Watson, index, for a complete list of Cardano's works owned by Dee. Unfortunately, only one has been recovered.

had inspired the work of philosophers such as Socrates, Plotinus, and Synesius, and he argued that guardian angels influenced his own natural philosophy. Cardano enjoyed visions and instructions from his personal angel while in prison and remained confident that his angel "was a spirit of power" because some of his visions were witnessed by others.[50] Cardano even credited his angel with helping him escape from captivity.

From Dee's library, his marginalia, and the discussions later held with angels about their natures and abilities, we can determine the basic features of his angelology. Essentially, he believed angels were composed of two substances: one active, the other potential.[51] Their duality of substance enabled them to participate with the divine and communicate with humanity without losing their intermediary place in the cosmic system. In accordance with their essential duality, angels were not limited by a material, corporeal form. Instead, they were spiritual and immortal.[52] Angelic spirituality became important to Dee's angel conversations because the angels were not limited to human methods of corporeal communication. According to Dee, the angels "have no organs or instruments apt for voyce: but are mere spirituall and nothing corporall: but . . . [they] have the powre and property fro[m] god [*sic*] to insinuate . . . [their] message or meaning to eare or eye . . . [so] that . . . they here [*sic*] and see . . . [the angels] sensibly."[53] The power to communicate through the imagination rather than speech was not the only supernatural ability enjoyed by the angels. They were also able to imitate divine thought processes, and their intellects were far superior to rational, human intellects.[54] Dee noted that the angelic intellect was greater and more powerful than the human, and their mental acumen was critical to his rationale for conversing with them.[55] Through their powerful intellects the angels had certain knowledge of the Book of Nature and,

50. Cardano, *Book of My Life*, p. 241.
51. See Dee's marks of emphasis in Julius Caesar Scaliger, *Exotericarum Exercitationum Liber Quintus Decimus, de Subtilitate, Ad Hieronymum Cardanum* (Paris, 1557), pp. 462–476, especially p. 467ᵛ. Roberts and Watson, #476.
52. Azalus, p. 19ᵛ.   53. Dee, *AWS* II: 330; see also Azalus, p. 19ᵛ.
54. See Dee's emphasis in his copy of the Dionysian corpus, Roberts and Watson, #975, pp. 13ᵛ–14ʳ [sig. B7ʳ]: "Sanctae itaque coelestium sprituum distinctiones, eis & que solum *sunt*, & qu[a]e sine rationis usu *viv*unt, & quae humana utu[n]tur *ratione* divinae *illius participationis dignitate & copia longe praecellunt*." Dionysius, "Celestial Hierarchy," p. 156.
55. Azalus, p. 18ᵛ. Dee noted that the "intellectus Angelicus" exceeds even the "intellectum sapie[n]tissimi hominis."

because they could see God, were able to see all aspects of the created world through God.[56] This did not mean that the angels knew the Deity completely, or knew about all divine causes and effects, but Dee believed that they did have a mastery over the natural and supernatural worlds.[57]

Such mastery was, by itself, insufficient reason for Dee's devotion to his angel conversations as a means for achieving natural philosophical aims. What raised Dee's hopes that the angels' knowledge would be communicated to him was his belief in guardian tutelary angels. A frequently employed trope in his angel conversations was the idea that he and Kelly were students in a "celestial school" with angelic "schoolmasters."[58] Dee's faith in tutelary angels was shared by others of the period. Cardano also believed that his best insights came from angelic guides, which he associated with the most subtle working of his intellect. First, Cardano believed he gained knowledge through the senses and observations. Second, he came to understand "higher things" by looking into their "beginnings." At this level, Cardano noted, he was "aided on many occasions by spiritual insight." The third, and highest, form of knowledge was "of things intangible and immaterial." Cardano arrived at this highest form of knowledge "wholly as a result of the ministrations of my attendant spirit."[59]

Because the emphasis on tutelary angelic guidance brought natural philosophy very close to revealed theology, Dee's move away from observation toward a revealed understanding of the Book of Nature has made problematic the status of the conversations as a form of natural philosophy. The importance and certainty of angelic revelations in matters pertaining to the natural world were, in Dee's time, the most significant features of his angelology. In his copy of Dionysius, Dee noted that angels were granted divine enlightenment and became responsible for passing arcane information to the lower, human levels of the cosmic hierarchy according to God's will.[60] Dee believed that the information

56. Dee's marginal notes in Azalus, p. 18ᵛ. In the chapter concerning the orders of angels Dee wrote: "Videntes Deum, vident o[mn]ia!"
57. Dee's emphasis in Azalus, p. 18ᵛ. "Vide[n]t etia[m] in Deo res, tanq[uam] in causa omnium primaria & efficie[n]te sed quia no[n] vident *totalitate[m] virtutis* divinae, q[ui]a hoc est solus Deus, non vident *omnia eius causata, seu o[mn]es effectus* eius" (Dee's emphasis).
58. *TFR*, pp. 102 and 367; Dee in Josten, pp. 236 and 248.
59. Cardano, *Book of my Life*, pp. 245–246.
60. Dee's emphasis in Dionysius, *Opera*, p. 14ʳ [sig. B7ʳ]: "Iccirco & prae omnibus *angelicum* cognomen merverunt, q[uae] cum in eas primo divin[a]e claritatis fulgor *emanet*, ipse postmodum in nos enu[n]ciationes illas sublimiores superne trancia[n]t." Dionysius, "Celestial Hierarchy," p. 157.

shared by the angels would include divine law, revelations of future events, prophecies, and doctrines. What is more, he was convinced that he would be transformed by this process, as would the world.[61] Once the purifying, illuminating, and perfecting light of God was mediated and transferred through the angels' revelations, Dee would be able to purify and perfect those around them.[62] The purified would then be able to receive further revelations, and turn to raise those beneath them in the cosmic hierarchy "toward *cooperation* with God."[63]

Dee faced a significant problem within his "celestial school": how to recognize the good angelic teachers from the demonic. He relied on visual cues and the angels' appearance for assistance.[64] He and Kelly always described even the most regular angelic participants – the angels Gabriel, Michael, Raphael, and Uriel – noting minute changes in their clothing or appearance. The appearance of angels was important in ancient and medieval angelology and in the Judeo-Christian tradition, where angels adopted visible, human forms.[65] Dionysius devoted an entire chapter of the "Celestial Hierarchy" to the ways in which angels appeared to humans, and Dee took note of the shapes, clothing, and attributes of the angels – from wheels of fire to flaming swords.[66] Yet the angels' appearance in a form legible to humans represented an abasement: "Angels abase themselves, to pleasure man by theyr instructions," Dee wrote, "when they tak[e] upon them, or use any sensible evidence of them selves or voyces, etc."[67]

Dee took to heart the warnings of scholars, including Trithemius, that no good angels ever appeared in the form of women. Dee's ambivalence about female angels caused problems within the "celestial school" when the female angels Madimi and Galuah appeared in the showstone. In order to "stop the mouths of others" who would dispute their trustworthiness, Dee questioned Galuah closely about whether good angels could appear in female form. Galuah responded that angels were neither male

61. Dee's emphasis in Dionysius, *Opera*, p. 14 ʳ⁻ᵛ [sig. B7ʳ⁻ᵛ]. Ibid. See also Dee's emphasis in Azalus, p. 18ᵛ: "*Ideo* per *Angelos* ad homina [*sic*] lex divina p[ro]cessit, o[mn]is *prophetia* o[mn]is *revelatio, atq[ue] doctrina*."

62. Dee emphasized "purgentur" and "p[u]rgent" in Dionysius, *Opera*, sig. B3ʳ. Dionysius, "Celestial Hierarchy, pp. 154–155.

63. Dee's emphasis, Dionysius, *Opera*, sig. B3ʳ⁻ᵛ. Dionysius, "Celestial Hierarchy," p. 155.

64. Dee, *AWS* II: 331; Harkness, "Shows in the Showstone," *passim*.

65. Azalus, p. 19ʳ. "Si vero *Angelus sub aliq[ua] visibili forma*, vel somno appeareat ho[min]i, vel loquat[ur] . . ." (Dee's emphasis).

66. Dee's emphasis in Dionysius, *Opera*, p. 48ᵛ–54ᵛ.

67. Dee, *AWS* II: 330.

nor female, but "take formes . . . according to the discreet and appliable [*sic*] will both of him [God], and of the thing wherein they are Administrators."[68] Though Trithemius "spake in respect of filthinesse" in regard to women, and separated "the dignity of the Soul of women from the excellency of man," Galuah reported that the abbot could not be more wrong. Galuah added that Wisdom was always "painted with a womans garment," and so it was appropriate for angels to appear as women, since their messages revealed divine wisdom. Dee's doubts were thus satisfied that the angels' forms were symbolic images created to assist frail human minds in understanding the mystical and profound truths they offered.

Dee's interest in the nuances of angelology demonstrates his efforts to master what he believed was a new and promising universal science. The angelic "schoolmasters" who taught him natural philosophy were an answer to his intellectual crisis of confidence, and the culmination of his earlier efforts to create a universal science capable of deciphering the Book of Nature. The angels offered Dee a role model, and a wider perspective on the decaying world, and gave him the tools to address the difficulties associated with practicing natural philosophy at a time when the natural world seemed unreliable and mutable. What is striking about Dee's studies in angelology was his ongoing fascination with the power of intermediaries, first evident in the *Propaedeumata aphoristica*, the *Monas hieroglyphica*, and the "Mathematical Preface." In Dee's angelology the intermediary communication that angels provided between the mundane and the divine was of paramount interest. Dee's next challenge was to conceive of a method to further his ambitions for angelic communication that would neither comprise his status as a natural philosopher nor violate the religious tenets of his time.

Dee's Method: Optics and the Power of Prayer

Dee's interest in angels and angelology was fostered by his belief that a more accurate assessment of the Book of Nature would result. In addition, he viewed his communication with angels as a new form of religious devotion in a period when religious practices were hotly contested. Like many religious practices in the early modern period, Dee's angel conversations could be seen as magic just as easily as they could be seen as either natural philosophy or religion, and it is this tangle of religious, magical, and natural philosophical ideas that has proven so problematic

68. *TFR*, p. 13.

to modern scholars. Dee's angelic cosmology combined religious and natural philosophical precepts into a way of looking at the world that offered hope of greater communication and wisdom. After all, the presence of angels in the world was of little use to a natural philosopher such as Dee if he could not communicate with them. To do so he enlisted additional religious and natural philosophical ideas that verged on the magical, specifically optics and the power of prayer.

Dee combined prayer and the magnification of divine influences through a crystal stone to facilitate his conversations with angels. His use of a crystal stone is the clearest example of the ways in which natural philosophical principles informed his practices. The showstone was the focal point of the conversations; it ordered the physical space and drew on the optics so central to Dee's natural philosophy. Understanding the properties of the showstone was, in Dee's mind, his supreme achievement, because its use produced tangible results in the form of new information and angelic communication. Dee's cosmology, as we have seen, recognized the secret potentials embedded in the Book of Nature as well as more manifest evidence of God's power. Nowhere is this duality more marked than in Dee's interest in the physics and metaphysics of light. As a result of these studies, he could relate his showstone and its use to the existence and behavior of visible and, more important, *invisible* rays. If, as Nicholas Clulee has persuasively argued, "Dee's natural philosophy was based on a theory of astral radiation," then his use of the showstone can be seen as the culmination of his natural philosophy, and quite possibly his "archemastrie."[69]

Although many occult practices, including divination, relied on stones or mirrors, Dee's utilization of the showstone was more optically sophisticated. In late medieval and early modern Europe stones were used in divination typically to locate stolen property, a practice decried by the church.[70] By the sixteenth century scrying had evolved into two interrelated traditions: catoptromancy, or the use of mirrors to see visions; and crystallomancy, or the use of crystal stones (such as Dee's showstone) to

---

69. Nicholas H. Clulee, "Astrology, Magic, and Optics: Facets of John Dee's Early Natural Philosophy," *Renaissance Quarterly* 30 (1977): 678. I have used the assertions of this early article to contradict Clulee's later belief that the angel conversations were influenced by Dee's natural philosophy but "cannot be considered as science or natural philosophy in its own right." See Clulee, p. 203.

70. Benjamin Goldberg, *The Mirror and Man* (Charlottesville: University of Virginia Press, 1985), pp. 12–13; Keith Thomas, *Religion and the Decline of Magic* (New York: Charles Scribner's Sons, 1971), p. 215.

see visions. One of the differences between catoptromancy and crystallomancy is that accounts of the latter almost always mention angels or other spirits, while scrying texts involving mirrors do not always refer to a supernatural agency.[71] Though catoptromancy and crystallomancy were the most common traditions, they do not represent the full spectrum of practices. Scryers could use cups of liquid to see visions (cyclicomancy), as well as rivers and ponds (hydromancy), and even the polished fingernails of virgins (onychomancy).[72]

Commonly, scryers used mirrors. Even the polished fingernail of a child or virgin was more popular than a crystal showstone like Dee's.[73] Most natural philosophers interested in the occult sciences during the early modern period, and specifically those interested in communicating with angels and spirits, tended to use crystal stones however. Simon Forman's book of magic, for example, included instructions for engraving crystals to facilitate the appearance of spirits and angels.[74] When crystal stones were used, the manuscripts often mentioned ritualistic preparations such as fumigations or anointing, which Dee did not include in the descriptions of his practices.[75] Yet Dee was well versed in occult traditions and owned works containing such rituals by Agrippa and Peter d'Abano. How and why did Dee's use of the crystal showstone differ from the early modern norm?

Dee's angel conversations employed the physics and metaphysics of light to lend the authority of natural philosophy and theology to traditional scrying practices and distance his use of the showstone from common forms of divination. The light metaphysics of Grosseteste and Roger Bacon informed Dee's earliest published works, the *Monas hieroglyphica* and the *Propaedeumata aphoristica*, as well as the later "Mathematical Preface" and angel conversations.[76] In the *Propaedeumata aphoristica* Dee described the metaphysical connection between light and certain "spiritual species" that were able to "flow from things both

71. Whitby, *AWS* I:76.   72.   Goldberg, *Mirror and Man*, p. 8.
73. See, for example, Cambridge University Library MS Ad. 3544, f. 27^r and f. 33^r. The Sloane collection of manuscripts at the British Library contains many accounts of occult activity involving crystal stones in five separate collections. See British Library MSS Sloane 3849, ff. 1–27; Sloane 2544, f. 59^v; Sloane 3846, ff. 61^v, 64^v, 148–151, 157–161; Sloane 3851, ff. 39–40, 50, 52, 92–109, 115; and Sloane 3884, f. 57^v–61.
74. Forman, "Magical Papers," British Library Add. MS 36674, f. 36^v.
75. Cambridge University Library MS Ad. 3544, f. 33^r.
76. See Chapter 2 for a complete discussion of these developments.

through light and without light."[77] This connection, along with the privileged position Dee accorded to light, forged a metaphysical connection between light, perfection, and spirituality. Though the mention of Dee's angel conversations may conjure in our minds the image of darkened rooms and the curtained trappings of a Victorian séance, the vast majority of angel conversations took place after sunrise and before sunset – although a brightly shining sun in England or northern Europe might have been too much to expect. In the earliest surviving conversation, the angel Anael instructed Dee to communicate with the angels on "the brightest day, When the Sonne shyneth: in the morning, fasting, begynne to pray. In the Sonne Set the stone."[78] Dee adhered to these instructions as best he could, as may be seen in his notations of the time of each conversation, as well as the date.

Although the crystal showstone facilitated communication with angels, it was not sufficient to ensure it. Dee's other safeguard was prayer. Prayer was a sanctioned method of communicating with the divine, and Dee's optical knowledge and his faith in the power of prayer worked together to provide him with both a theory and a practice for rising above his limitations as a flawed human being in a decaying and corrupt world. Most important, however, prayer offered Dee a method of communication with the occult forces of the Book of Nature that did not – technically, at least – resort to invocation or the binding of spirits. Bishop John Fisher, one of the founders of Dee's undergraduate college, described the power of prayer as a

> golden rope of chaine lett downe from heaven, by which we endeavour to draw God to us whereas we are more truly drawne to him. . . . This rope of golden Chaine holy S. Dionisius calleth prayer. . . . Lastly what other thing doth it, but elevates the mind above all things created, soe that att last it is made one spirit with God. . . . [79]

Prayer was a way for people to rise to higher intellectual levels and to climb the cosmic ladder. Dee, however, took traditional theological arguments about the power of prayer and the existence of angels a step further with his conviction that communication with angels was not limited to mystics and theologians; it was also the prerogative of natural philosophers well versed in the Book of Nature.

77. Dee, *PA*, pp. 127–129.   78. Dee, *AWS* II:15.

79. John Fisher, *A Treatise of Prayer and the Fruits and Manner of Prayer*, pp. 19–20 and 22; quoted and discussed in L. Kelly Faye, *Prayer in Sixteenth-Century England* (Gainesville: University of Florida Press, 1966), pp. 29–30.

Dee's "extra step," his belief that communication with angels represented yet another tool in the workshop of a natural philosopher, was
not entirely idiosyncratic. Other natural philosophers interested in deciphering the Book of Nature were also interested in communication with
the spiritual levels of the cosmos. Where they differed from Dee was in
their reliance on binding spells rather than prayer to command spirits.
For example, Thomas Allen (d. 1633), one of early modern England's
greatest bibliophiles and collectors, owned a sixteenth-century manuscript that described how to invoke and bind simple spirits.[80] Another
early modern manuscript collection included instructions for conjuring
angels with the aid of a crystal and invocations and requiring them to
perform spiritual "experiments." An interesting feature of this manuscript is the account of a "true experiment proved in Cambridge anno
1557 of 3 spirits to [be] don[e] in a chamber, whose names are Durus,
Artus, Aebadel."[81] Dee spent considerable time at Cambridge University,
and this manuscript suggests the intriguing possibility of a link between
these earlier, anonymous spirit experiments and Dee's later decision to
pray for angelic guidance.

Calling upon God through prayer and summoning spiritual agencies
through magical invocation are technically distinct, although the passage
of time has made them less so. In prayer, the practitioner subjects himself
to the will of God. In magical invocation, on the other hand, the practitioner subverts the hierarchical arrangement of the cosmos by asserting
his or her will over a spirit and, through a subsequent binding spell,
controlling a spirit's actions. Though prayer and invocation might have
the same roots in early religious practices, by the sixteenth century the
distinction between them was pronounced. Dee would have argued that
his conversations with angels involved no hint of invocation but employed only prayer and his devout subjection to God's will and purposes.

80. Allen is a complex figure, who may have employed Edward Kelly before
    he entered Dee's service as a scryer. The connections between Allen and
    Dee do not end there. Count Albert Laski invited him to Poland in 1583 –
    when Laski extended a similar invitation to Dee – but Allen chose to
    remain in England and collaborate on intellectual projects with Henry
    Percy, the "Wizard Earl" of Northumberland. See Bodleian Library e. Mus.
    MS 238, passim. There are no marks on this manuscript to indicate that it
    ever passed through Dee's hands. See Andrew G. Watson, "Thomas Allen
    of Oxford and His Manuscripts," in *Medieval Scribes, Manuscripts, &*
    Libraries. Essays Presented to N. R. Ker (London: Scolar Press, 1978),
    pp. 279–316; Roberts and Watson, pp. 58 and 65.
81. Bodleian Library Rawlinson MS D253, f. 50ʳ and passim.

Dee had been sensitive to charges of conjuring and invocation since 1547, when he was alleged to have commanded spiritual agencies in a production of Aristophanes' *Pax* at Cambridge University. For the production Dee engineered a beetle that miraculously flew through the air "to Jupiter's pallace, with a man and his basket of victualls on his back."[82] The sensation caused by the flying beetle was sufficient to start the first whispers that Dee had achieved his special effect, not by mechanics, but by magic. Years later, Dee was still lamenting the "vaine reportes spread abroad of the meanes how that [beetle] was effected."[83]

The tension between prayer and invocation is linked to the larger problem of what constituted magic in the early modern period. Defining magic can be difficult, because neither early modern authors nor modern scholars can agree on how words like *conjuring, magic,* and *invocation* should be used.[84] Many modern scholars, following the lead of Frances Yates, see Marsilio Ficino's revival of the Hermetic corpus as the pivotal moment in the development of early modern magic.[85] Ficino's interest in the ideas attributed to the pre-Christian philosopher Hermes Trismegistus, they argue, helped to foster a conception of magic embedded both in the natural properties of the world and in the supernatural powers linked to those properties through sympathetic relationships. In the early modern period, however, the ideas of Hermes were pagan, even if he was attractive to humanists because of his great antiquity. While many, including Ficino, argued that the Hermetic corpus could be read and utilized by orthodox Christians because it contained prophetic references to Christ, other passages (such as fumigations and the activation of statues through astral and sympathetic magic) were contrary to church doctrines. Later, Yates emphasized her initial premise – that Hermeticism alone did not account for the early modern interest in the occult – but

---

82. Dee, CR, pp. 5–6.   83. Dee, CR, pp. 5–6.
84. Brian P. Copenhaver, "Natural Magic, Hermeticism and Occultism in Early Modern Science," in *Reappraisals of the Scientific Revolution*, ed. David Lindberg and Robert Westman (New York: Cambridge University Press, 1990), pp. 261–301. Instead of using specific terms like "hermeticism," Copenhaver suggests the alternative term "occultist" to describe early modern interest in the workings of the unseen forces of nature as it was manifested in demonology, magic, astrology, and other disciplines (p. 289).
85. See Frances A. Yates, *The Occult Philosophy*, p. 1, and *Giordano Bruno and the Hermetic Tradition* (London: Routledge and Kegan Paul, 1964), pp. 62–83.

her efforts to return to a more varied conception of magic are often overlooked in favor of a purely Hermetic viewpoint.[86] In contemporary scholarship, Hermeticism is still considered the prime mover of early modern occultism and has become synonymous with magic as a result – an inaccuracy that undermines our analysis of early modern natural philosophy.

Whenever scholars have tried to classify Dee's angel conversations, they have quickly discovered that no single category such as Hermeticism can contain them. The conversations were explicitly separated from the Hermetic tradition, as the angels told Dee, in an allusion to Moses' contest with Pharoah's magicians in Exodus 7:9–13, that the "Arts of the Egyptian Magicians" seemed to be powerful but were not.[87] Nor do the angel conversations fall easily under the constraints of "natural magic" in which properties of the natural world are manipulated and enhanced, as in the work of one of its leading proponents, Giovanni Battista della Porta.[88] Dee never contacted the devil or made any sort of pact exchanging his soul for information, so it is impossible to consider the angel conversations a form of black magic. Even demonic or spiritual magic, where a good or bad spirit is invoked, bound, and forced to do a magician's will, fails to characterize Dee's angel conversations. Nor can they be classified as ceremonial or "white" magic as Wayne Shumaker defines them, as Dee did not use the vital process of ritualistic invocation.[89]

If today's scholars are confused about the precise nature of magic in the early modern period, they are not alone; it is clear that Dee and his contemporaries were not entirely sure what magic, religion, or natural philosophy were either. Though most individuals accused of being magicians were vehement about their innocence, early modern definitions of magic were diverse, incommensurable with each other, and internally incoherent. An example from the work of Francesco Giuntini (1523–1590), who lived in the French city of Lyons during Dee's visit there, illustrates the point. In his *Speculum astronomiae* (1573) Giuntini discussed the possibility that magic and necromancy might well be branches

86. Yates, *The Occult Philosophy in the Elizabethan Age* (London: Routledge and Kegan Paul, 1979), especially pp. 1–6.
87. *TFR,* p. 116.
88. See John Baptista della Porta, *Natural Magick* (London, 1658), for an early modern English translation of the text.
89. Wayne Shumaker, *The Occult Sciences in the Renaissance* (Los Angeles: University of California Press, 1972), pp. 108–109.

of natural philosophy subordinated to astrology.[90] Giuntini based this opinion on the following argument: "[in magic and necromancy] demons operate nothing except by natural application of active forces to the appropriate and proportionate passive objects, which is the work of nature, and this at determined times and under determined signs, which is astrology.[91] Though logically flawless, Giuntini's opinion was not shared by most of his contemporaries – especially in regard to necromancy, or conjuring with the dead, which remained the most objectionable form of magic in the early modern period. Giuntini went on to argue that magical abilities, if acquired through study or from a good angel, were completely unobjectionable no matter what their form or function.[92] Giuntini's ideas might have been warmly received, especially by someone like Dee, but magic remained in a liminal and problematic place in early modern systems of knowledge.

Dee and his contemporaries drew fine – sometimes very fine – distinctions between magic, religion, and natural philosophy in an attempt to relieve the tensions among them. Often, authors of the period made a particular point of excepting communication with angels from other forms of magic, as did Giuntini. Prior to Giuntini, the positive results that might be had from communication with angels were suggested by Agrippa. In *De occulta philosophia* Agrippa outlined a system of occult beliefs and practices that was at once natural, spiritual, magical, and deeply religious. His system relied heavily on the fusion of mathematics and language present in the cabala, which became a chief element in Dee's angel conversations. But Agrippa placed a far greater emphasis on the wonder-working potential of names and their use in summoning spirits and angels. In addition, Agrippa was interested in the magical practices endorsed by Ficino's Hermeticism, such as fumigations and the use of magical sacrifices to draw and enhance occult powers in the natural world.[93] These practices were absent from Dee's angel conversations, too, and their absence suggests that he was adapting the angelic magic suggested by Agrippa and Giuntini in favor of a more streamlined approach that would defy fewer orthodox religious practices. This is evident in the absence of invocations and compulsions in Dee's angel

90. Dee owned three of Giuntini's works, including the *Speculum astronomia* (Paris, 1573), which he took with him to central Europe; it has not been recovered. See Roberts and Watson, #470, #739, and #1883.
91. Giuntini, ff. 45ᵛ–47ʳ, quoted and discussed in Thorndike, VI:132.
92. Giuntini, f. 47ᵛ, quoted and discussed in Thorndike, VI: 132.
93. Agrippa, *DOP*, passim; Shumaker, *Occult Sciences*, p. 155.

conversations, both of which were prevalent in other spiritual and demonic practices, in favor of humble prayer and petitions. This absence led Clulee to argue persuasively that Dee did not see the angel conversations as "a type of magic but as a variety of religious experience sanctioned by the scriptural records of others to whom God or his angels imparted special illumination."[94]

Dee used prayer to facilitate the angel conversations rather than spells and conjuring, but prayer could never be completely divorced from occult practices.[95] Sir Walter Raleigh drew on Plato to argue that the "art of magic is the art of worshipping God."[96] Dee emphasized in his copy of Azalus's encyclopedia that with the use of techniques such as prayer it was possible to communicate with angels without committing the sin of conjuring.[97] He relied on a prayer which *requested* that God send angels to communicate with him. There is no sense in the angel diaries that the angels were called to do Dee's bidding; they were present at the command of God to do His will. Even Meric Casaubon acknowledged, in his preface to *A True and Faithful Relation*, that Dee "could express himself very fluently and earnestly in Prayer," and that Dee's complete faith in God was impeccable – it was the godliness of his scryers that was in doubt.[98]

The role of prayer in Dee's angel conversations reflected the place prayer occupied in post-Reformation Europe, where it was an important feature of theology and daily life. Through prayer and supplications, wishes were answered, requests for assistance were granted, and divine support was obtained – as long as God willed it. According to Keith Thomas, "there was no benefit which the pious Christian might not obtain by praying for it."[99] Communicating with the divine was a tricky business, however, and the difficulties of achieving perfect communication with God were often acknowledged. Recently, Cynthia Garrett revealed that prayer manuals in seventeenth-century England contained "a complex theory of prayer which acknowledges, at times even embraces,

94. Clulee, p. 206.
95. Richard Kieckhefer, *Magic in the Middle Ages* (New York: Cambridge University, Press, 1990), pp. 69–75; Keith Thomas, *Religion and the Decline of Magic* (New York: Charles Scribner's Sons, 1971), pp. 41–42.
96. Walter Ralegh, *The History of the World* (London, 1614), I:xi: 2, p. 201.
97. Azalus, p. 19ʳ. "*Arsque notoria* de appare[n]tibus figuris bonoru[m] angeloru[m], q[ua]m devote *orationibus, & sine peccato* mortali advocantur. . . ."
98. Meric Casaubon, "The Preface," in *TFR*, sig. Eʳ⁻ᵛ.
99. Thomas, *Religion and the Decline of Magic*, p. 113.

the contingent and imperfect nature of communication with the divine."[100] While communicating with God could be fraught with tensions and even danger, it was nonetheless a positive experience.

Prayers were used throughout the Middle Ages, yet medieval prayers often relied on an intermediary – a saint, the Virgin Mary, or one of the angels – to facilitate the petition. In Dee's time, however, Reformation theologians placed a new emphasis on the ways in which prayer could foster an "unmediated relationship between the individual and God."[101] Dee showed himself, in this respect at least, very much a Protestant, for he did not direct his prayers to intercessory agents, such as the angels or the saints.[102] He communicated with God *directly*, and asked the Deity to send angels, who would then function as intermediaries in the transmission of divine knowledge. Simon Forman, in a prayer dated 1567, placed a similar emphasis on contacting angels to obtain knowledge through a Protestant supplication to God rather than an intercessory agency.

> O god of Aungells, god of Arch aungells; god of Patriarches, god of Prophetts, god of us Sinners; O Lord be my helpe, that this my worke may proceede in good tyme, to thy glorie O God; and so learninge, and not aut else, that I would this day have. O my God be in my tounge, that I may glorifie the[e] in all workes, Amen.[103]

After the angels revealed designs for the manufacture of wax seals, Forman's prayers continued with a request for protection against evil and the disclosure of information.

> Let not [*sic*] evyll spyritt enter my mynde o God, nor nothinge else but all to thy glorie o God; for learninge is all my desior, lord thou knowest, even as yt was to thy servaunte Solomon; O lorde sende me som[m]e of his good hidden worke, that hath not been revealed to noe man. Then for that cause I desier the[e] O god to send at mee, that in these dier laste daies yt may be knowen.[104]

The importance of God's "good hidden worke" to Forman, as in the case of Dee, was linked to his belief that the "dier last daies" of the world were at hand and that natural philosophers could play a role in the reordering of nature.

After the Reformation, the Protestant belief that the end of the world was approaching placed increased importance on prayer, since the Bible

---

100. Cynthia Garrett, "The Rhetoric of Supplication: Prayer Theory in Seventeenth-Century England," *Renaissance Quarterly* 46 (1993): 329.

101. Ibid., p. 329.   102. Ibid., p. 333.

103. Simon Forman, "Magical Papers," British Library Add. MS 36674, ff. 47ʳ–56ᵛ, f. 47ʳ.

104. Ibid., f. 46ʳ.

suggested God would communicate with the elect as the Day of Judgment approached. Catholic theologians followed the example of their Protestant adversaries as apocalyptic fervor increased, and by the end of the sixteenth century most of Europe was trying to communicate more completely with God. Their attempts included more precise rituals on the part of Catholics and more precise language on the part of Protestants.[105] Reformed prayer was necessary because, like everything else in the Book of Nature, prayer had deteriorated since the time when the descendants of Adam and Eve first petitioned God. Still, as the medieval cleric and natural philosopher Roger Bacon noted in his *Epistolae de secretis operibus artis et naturae, et de nullitate magiae*, some prayers retained their original power. These prayers, "instituted of old by men of truth, especially those ordained to God and to the angels . . . are able to retain their original virtue."[106]

Praying correctly was only one of the problems facing a devout person in the sixteenth century. In addition, Dee and his contemporaries worried about their ability to decipher God's response. Theologians agreed that God always answered human prayers, though the response could be negative. The faulty interpretation of those responses was a real problem, however; humans often failed to "read God" correctly, as His responses were protected by layers of encoding and symbolism. Like many of the "imperfections" with which humans had to suffer, the unsatisfactory state of communication between heaven and earth stemmed from original sin. Because of the moral Fall, once perfect lines of communication between heaven and earth had deteriorated and God had to reveal His answers to prayers in signs rather than speech.[107] Most theologians believed that God's symbolic messages were particularly evident in the Book of Nature, where they took the form of comets and other marvels.[108]

Dee's angel diaries indicate that prayer was a major part of his strategy

105. Garrett, "Rhetonic of Supplication," p. 355.
106. Roger Bacon, *Roger Bacon's Letter concerning the Marvelous Power of Art and Nature and concerning the Nullity of Magic*, trans. Tenney L. Davis (Easton, PA: Chemical Publishing Company, 1923), p. 18. An earlier translation, purportedly transcribed from Dee's copy, is *Frier Bacon His Discovery of the Miracles of Art, Nature, and Magick. Faithfully translated out of Dr. Dees own Copy by T. M. and never before in English* (London, 1659).
107. Garrett, "Rhetoric of Supplication," p. 353.
108. See, especially, Bryan Ball, *A Great Expectation: Eschatalogical Thought in English Protestantism to 1660*, Studies in the History of Christian Thought, vol. 12 (Leiden: E. J. Brill, 1975), passim.

for communicating with the divine. In one conversation, an angel spoke at length about the virtues of prayer. "Prayer is the Key, sanctified by the Holy Ghost, which openeth the way unto God," the angel told Dee.[109] Michael told Dee that, without invoking the name of God, angels could do nothing in the terrestrial world, and that *"The key of Prayer openeth all things."*[110]Most of Dee's diary entries opened with references to prayer, which preceded any references to visions in the showstone. Dee and his associates were instructed to share a "Conjunction of myndes in prayer" to communicate with God and receive angelic revelations.[111] Often the prayers were traditional: Dee frequently used one or more of the Psalms in the angel conversations, for example, and was told by the angel Uriel to use the seven penitential psalms to communicate with Michael.[112] At other times, the prayers were more personal: Dee wrote lengthy original prayers for himself and others. On one occasion he recorded how he was writing traditionally Catholic, intercessionary "prayers to good Angels . . . for A.[lbert] Lasky," when an angel suddenly appeared in the stone.[113]

Prayer was the most overtly religious feature of Dee's angel conversations, but it was by no means the only practice with religious connotations. In diary notes dated 22 December 1581, Dee recounted the angel Anael's instructions to fast, abstain from sex and gluttony, be clean and neatly groomed, say their prayers seven times facing each of the four cardinal directions, and be secretive about the angel conversations as evidence of their devotion.[114] In addition, Dee was told not to engage in angel conversations on Sundays, to pray continually, wait for the time when God would "stir" him to conversation, and consider astrological influences on the conversations. "Work in the Sunshine, and from the change of the [moon] to the 14 day after," the angels told Dee, and "in increasing hours, which are from [sun] rise to Noone . . . let the Sun be well placed, & a good planet raigning."[115]

Some of these instructions were similar to ideas found in the spurious fourth book of Agrippa's *De occulta philosophia*. In this work (believed authentic during Dee's lifetime), Agrippa pointed out that it was important for those engaged in communication with spirits to have been "religiously disposed for many days to such a mystery." Ideally, the partic-

109. *TFR,* p. 384.　110. Dee, *AWS* II:83; Dee's emphasis.

111. Ibid., 20.　112. Ibid.　113. *TFR,* p. 24.

114. Dee, *AWS* II:14–16. Elias Ashmole extracted the information concerning ritual and procedures and collected them under the heading "Notes for Practise." See Bodleian Library MS Ashmole 1790, f. 39ʳ.

115. Bodleian Library Ashmole MS 1790, f. 39ʳ.

ipants should be confessed of their sins, contrite of heart, and washed with holy water. They should be chaste and abstinent, and not preoccupied with secular business. In addition, participants should have fasted as much as possible for forty days, or one lunation.[116] In the third book of *De occulta philosophia*, Agrippa emphasized that successful occult practices were based on a strong sense of faith rather than adherence to a grimoire or set of magical precepts. Faith distanced the natural philosopher from the perils of the world, and Agrippa reminded his readers that trusting too much in nature rather than in God was a sure way to communicate with bad demons instead of good angels.[117]

In 1586, when Dee was in Prague, the angels modified the practices associated with the conversations. The changes further complicate the portrait we have of Dee's angel conversations, because the new prayers relied on Roman Catholic liturgical forms that would have been acceptable in Prague. In later descriptions of the angel conversations, Dee emphasized his role as the "common advocate" for the group, serving his associates as a Catholic priest would serve a congregation, voicing the wishes and desires of the entire group before God. After crossing himself while saying "In the name of the Father, of the Son, and of the Holy Spirit," Dee recited the Lord's Prayer. He continued with the following prayer:

> Almighty, Sempiternal, True and Living God, send out Thy light and Thy truth which may guide us safely to Holy Mount Sion and to Thy celestial tabernacles where we may praise and glorify you eternally and for ever and ever. Amen.[118]

Dee then silently repeated the Sixty-seventh Psalm three times:

> God be merciful unto us, and bless us: and cause his face to shine upon us; Selah.
> That thy way may be known up on earth, thy saving health among all nations.
> Let the people praise thee, O God; let all the people praise thee.
> O let the nations be glad and sing for joy: for thou shalt judge the people righteously and govern the nations upon earth. Selah.
> Let the people praise thee, O God; let all the people praise thee.

116. Henry Cornelius Agrippa, *His Fourth Book of Occult Philosophy; Of Geomancy; Magical Elements of Peter De Abano; Astronomical Geomancy; the Nature of Spirits; Arbatel of Magick*, trans. Robert Turner (London, 1655), pp. 59–60. Dee owned a copy of the 1559 edition, Roberts and Watson, #743, which he took with him to central Europe in 1583. It has not been recovered.

117. Quoted in Shumaker, *The Occult Sciences*, p. 146.

118. Dee in Josten, pp. 240–241.

> Then shall the earth yield up her increase; and God, even our own
>   God, shall bless us.
> God shall bless us; and all the ends of the earth shall fear Him.

While Dee silently prayed, the rest of the participants waited for the angels' revelations.

In Dee's later conversations the tension between Catholic and Protestant traditions indicates his belief that the angel conversations signaled a new program of religious reform that would unite the religious factions of Europe. Like the proponents of a universal science, the advocates of "universal religion" hoped that a code of faith might be found, constructed, or reconstructed to end the uneasy religious climate of the Reformation era. Of special concern was the conversion of the Jews to this new universal religion, a much anticipated sign that the End of Days was approaching. The religious sect known as the Familists were one of the most outspoken groups to promote universal religion, and Dee might have come into contact with Familist ideas during his residence in Antwerp. Many of his friends there, including Ortelius, Mercator, the publisher Plantin, and Gemma Frisius, were associated with Familism, whose main tenets were religious toleration, outward conformity to state religions, and a deep spiritual adherence to the teachings of the religion's founder, Hans Niclaes.[119]

119. The literature on Familism, or the "Family of Love" as it was known in England, is extensive. Many valuable insights into the connections between Familism and the leading proponents of literary and intellectual movements in the Elizabethan period can be found in J. A. Van Dorsten, *The Radical Arts: First Decade of an Elizabethan Renaissance* (London: Oxford University Press, 1970). In a more general vein, Alastair Hamilton, *The Family of Love* (Cambridge: James Clarke and Company, 1981), discusses Familism and its possible influence on Dee (passim, especially p. 113). Jean Dietz Moss has investigated the important early years of the Familist movement, as well as its English manifestations, in *"Godded with God": Hendrik Niclaes and His Family of Love*, Transactions of the American Philosophical Society, vol. 71 (Philadelphia: The American Philosophical Society, 1981), and "The Family of Love and English Critics," *Sixteenth Century Journal* 6 (1975): 33–52. The most recent book on the subject is Christopher Marsh, *The Family of Love in English Society, 1550–1630* (Cambridge: Cambridge University Press, 1993). Three studies of important individuals associated with the religion (Benito Arias Montano, Guillaume Postel, and Christopher Plantin) also contain useful information. See B. Rekers, *Benito Arias Montano (1527–1598)* (London: Warburg Institute, 1972); Marion Leathers Kuntz, *Guillaume Postel, Prophet of the Restitution of All Things: His Life and Thought* (The Hague: Martinus Nijhoff Publishers, 1981); and Leon Voet, *The Golden*

In summary, Dee would not have considered his angel conversations a "magical" practice. Instead, like many of his contemporaries interested in the occult properties of nature, he was devoted to living a properly religious life. It is true that he would not have satisfied a strict early modern criterion for orthodoxy – his interest in the cabala alone would have raised eyebrows – for he had a broader conception of what devotion to God involved. Dee, a Christian natural philosopher, read the Book of Nature so that he could glorify God. His predecessor Agrippa had integrated natural philosophical and religious interests even further, seeing religion "as one natural force among others, though the most elevated of those forces."[120] Dee's last universal science, his angel conversations, was thus more than an attempt to provide a unified basis for natural philosophy. It sought to unify and make coherent all religious beliefs, natural knowledge, and ancient theory.

Compasses: The History of the House of Plantin Moretus, vols. 1–2 (New York: Abner Schram, 1969), especially 1: 23–26.

120. Paola Zambelli, "Magic and Radical Reformation in Agrippa of Nettesheim," *Journal of the Warburg and Courtauld Institutes* 39 (1976): 85.

PART II

Revelations

Such will be the old age of the world: irreverence, disorder, disregard for everything good. When all this comes to pass . . . he will restore the world to its beauty of old so that the world itself will again seem deserving of worship and wonder. . . . And this will be the geniture of the world: a reformation of all good things and a restitution, most holy and most reverent, of nature itself, reordered in the course of time. . . .

(—*Asclepius* 26 from the translation by Brian Copenhaver)

4

"Then Commeth the Ende"

Apocalypse, Natural Philosophy, and the Angel Conversations

John Dee's plan to converse with angels and gain insights into the Book of Nature emerged from his particular intellectual interests, which he pursued through decades of scholarship and inquiry. But the angel conversations also reflected the cosmology of his time, especially the late sixteenth-century conviction that the end of the world was at hand. When Dee began contacting the angels around 1569, European scholars and theologians believed they were caught up in the midst of events prophesied in the Bible's Book of Revelations that would ultimately turn the final page of the Book of Nature.[1] Citing the decay of the earth and the heavens, the reformation of the church, and a plethora of strange natural occurrences, Dee's contemporaries lamented the final days while looking forward to the restitution of peace, plenty, and

1. Studies of early modern apocalypticism include: Robin Bruce Barnes, *Prophecy and Gnosis: Apocalypticism in the Wake of the Lutheran Reformation* (Stanford, CA: Stanford University Press, 1988); Richard Bauckham, ed., *Tudor Apocalypse: Sixteenth-Century Apocalypticism, Millenarianism, and the English Revolution* (Oxford: Sutton Courtenay Press, 1978); Frank L. Borchardt, *Doomsday Speculation as a Strategy of Persuasion: A Study of Apocalypticism as Rhetoric*, Studies in Comparative Religion, vol. 4 (Lewiston, NY: Edward Mellen, 1990); Paul Christianson, *Reformers and Babylon: English Apocalyptic Visions from the Reformation to the Eve of the Civil War* (Toronto: University of Toronto Press, 1978); Katharine R. Firth, *The Apocalyptic Tradition in Reformation Britain 1530–1645* (Oxford: Oxford University Press, 1979); Walter Klaassen, *Living at the End of the Ages: Apocalyptic Expectation in the Radical Reformation* (Lanham, MD: University Press of America, 1992); Rodney Lawrence Petersen, *Preaching in the Last Days: The Theme of "Two Witnesses" in the Sixteenth and Seventeenth Centuries* (Oxford: Oxford University Press, 1993). For an excellent overview of the antecedents of early modern apocalypticism, see Christopher Rowland, *The Open Heaven. A Study of Apocalyptic in Judaism and Early Christianity* (1982) (London: SPCK, 1985).

prosperity. The years for which the records of Dee's conversations survive were periods of heightened eschatological fervor: in 1572 a new star had been observed in the constellation of Cassiopeia; in 1577 Dee spent three days at Windsor Castle advising the queen about the significance of a comet; at Easter in 1580 an earthquake rocked London; and in 1583 a grand conjunction was predicted in the constellation Aries.[2] The grand conjunction, which occurred just as Dee left for Prague, was perhaps the most significant and widely publicized sign. When it was accompanied by a particularly bright comet, the final days seemed assured.

Although earthly corruptions were further proof of the decay of the Book of Nature, Dee's contemporaries found the celestial changes more alarming. Discrepancies between ancient cosmological calculations and sixteenth-century studies of the heavens provided further evidence of the end of time. In 1580 Francis Shakelton consulted natural philosophers and discovered that "the constitution of the celestiall worlde, is not the same that it hath been in tymes paste, for so much as the Sunne, is not so farre distant from us now, as it hath been heretofore."[3] Shakelton went on to ask, "If there be so greate alteration in the superior worlde, what shall wee saie of the inferiour?"[4]

Shakelton believed that the disturbing changes in the heavens pointed to an even greater mutability and instability in the imperfect sublunar world, where "every part thereof doeth feele some debilitie and weakenesse." Human corruption was proceeding apace along with nature's alarming deterioration. Andrew Golding confirmed that "mannes nature [is] growing dayly more and more into decay with the perishing worlde nowe hasting to his ende, is more subiecte too corruption, and less gyven too Godlynesse and vertue than ever it was."[5] Learning and knowledge were also on the decline, which Dutch commentator Sheltco à Geveren saw as another "argument of the worldes consummation," since "al good Arts and learning, have these few yeares been so con-

2. For an analysis of Dee's interpretation of the comet, see Robert William Barone, "The Reputation of John Dee: A Critical Appraisal" (Ph.D diss., Ohio State University, 1989), p. 51.
3. Francis Shakelton, *A Blazyng Starre or burnyng beacon* ... (London, 1580), sig. Aiiii^v–Av^r.
4. Ibid., sig. Av^r.
5. Golding, quoted in Ernest Lee Tuveson, *Millennium and Utopia: A Study in the Background of the Idea of Progress* (Berkeley: University of California Press, 1949), p. 32.

temned, and Universities and schooles and scholasticall discipline . . . almost in every place come to decay."[6] It was in this world of anxiety and anticipation that Dee's angel conversations took place.

By the time eschatological fervor had reached its highest pitch, Dee was already in despair over his inability to practice natural philosophy in a corrupted world. The events of the 1570s and 1580s did little to alleviate his concern. Though a natural philosopher's interest in the end of the world may seem puzzling to us, Robin Barnes points out that apocalypticism and natural philosophy were often linked in the early modern period because people believed that both nature and the *end* of nature were instrumental in God's plan for the cosmos. Barnes does not find it surprising, therefore, that scholars interested in establishing a universal science and reviving the traditions of ancient wisdom held eschatological and prophetic beliefs.[7] Dee's interest in the end of the world was evident in both his library and his published works.

Dee purchased nine works on the new star of 1572 and acquired the printed collection of letters exchanged by fellow Englishmen Gabriel Harvey and Edmund Spenser on the Easter earthquake of 1580.[8] He was already aware, through the works of Regiomontanus, that something cataclysmic was bound to happen at the time of the grand conjunction.[9]

6. Sheltco à Geveren, *Of the ende of this world* (London, 1577), sig. Diii^r. Dee's copy of this work is Roberts and Watson, #1727.
7. Barnes, *Prophecy and Gnosis*, pp. 26–27.
8. Dee's library books concerning the new star were: Marcus Manilius, *Astronomica* (Basel, 1533), Roberts and Watson, #251, which contains Dee's annotations on the new star; a work by Dee's landlord in Prague, Tadeas Hájek, *Dialexis de nova stella* (Frankfurt, 1574), Roberts and Watson, #438; Theodorus Graminaei, *Erklerung oder Auszlegung eines Cometen* (Cologne, 1573), Roberts and Watson, #703; Jerónimo Muñoz, *Du nouveau comète de l'an 1572* (Paris, 1574), Roberts and Watson, #842; Georg Busch, *Die andere Beschreibung vondem Cometen welcher in 1572 Iar erschienen* (1573), Roberts and Watson, #1291; two copies of Cornelius Gemma and Guillaume Postel, *De peregrina stella 1572* (Basel 1572/1573), Roberts and Watson, #1876; Cornelio Frangipani, *Sopra la stella dell'anno 1572* (Venice, 1573), Roberts and Watson, #2137; Augustus, Elector of Saxony et al., *De publica poenitentia . . . exorta peregrina stella 1572* (1578), Roberts and Watson, #2217; Joannes Sommerus, *Refutatio scripti Petri Carolii editi Wittebergae* (Cracow, 1582), Roberts and Watson, #D20; for the Harvey-Spenser letters, see Roberts and Watson, #1720.
9. For Dee's copies of Regiomontanus, see Roberts and Watson, #85/B310, #559, #D15, #9/B186.

Dee's *Propaedeumata aphoristica,* his *Monas hieroglyphica,* and his "Mathematical Preface" all attempted to reform and redirect natural philosophy through a new universal science, but they fell short of reaching the top rungs of the ladder to the heavens.

Only angelic revelations could begin to satisfy Dee's hopes for attaining certain knowledge of the Book of Nature. Those conversations included detailed information about the unfolding of the *eschaton* or final days of the earth, including symbolic and verbal information about God's plan for the dissolution and restitution of the cosmos. The angels revealed two new natural philosophical tools for Dee's use: an exegetical tool (the cabala of nature, discussed in Chapter 5) and a redemptive tool (the medicine of God, discussed in Chapter 6). Together, this information provided Dee with the natural philosophy of the future when natural philosophers would no longer need to struggle through textual explanations of the Book of Nature, search through ancient authors for answers to their questions, or try to close the distance between humanity and divinity by observing the stars in the heavens. Most (if not all) of the insights into nature gained after the Fall and before its restitution were virtually useless anyway, except as a yardstick with which to measure how far things had deteriorated. When the heavenly Jerusalem descended and the world was restored to perfection, the angels told Dee he would *know* – in the same way as they did – about every character found in the Book of Nature.

Before the world was perfected, however, Dee had other tasks to perform. First, he was asked to interpret the decay of nature based on available natural signs. Aided by the angels' revelations and tools, Dee's interpretations would far surpass all others. This information would enable Dee to act as a prophet of the final days, spreading revelations among the people of the world. Gifted with divine revelation, Dee believed himself to be invested with special responsibilities of communication he shared with the great biblical prophets Elias, Enoch, and John the Divine. Communicating his angelic revelations to a wider audience was one of Dee's more problematic responsibilities, but he took it seriously and turned his attention to the angels for further information about the End of Days.

Interpreting the End of Days

Today we tend to group all beliefs that the world is ending under the rubric of "apocalypticism." In the early modern period, however, apocalyptic expression fell into several different categories: apocalypse, prophecy, and eschatology. Neither the interest in the final days of the earth nor the response to that interest was homogeneous. While some speculation and divination was tolerated, other opinions met with considerable resistance. Still, the belief that the end of the world was approaching continued to flourish well into the seventeenth century, where it took on a new cast within shifting social, cultural, religious, and economic contexts. The lines dividing apocalypticism, prophecy, and eschatology were never so clearly demarcated as is suggested here, but it is important to distinguish among them because the distinctions can help us to understand why Dee was initially "well received" by monarchs and church officials yet later turned away when his angelic revelations became increasingly apocalyptic.

An apocalypse was a particular genre of speculation about the final days whose authors claimed to have received a direct revelation from a divine agency – either God, an angel, or a saint.[10] Apocalypses were usually communicated to the world through prophets, as in the case of Girolamo Savonarola.[11] Prophecy was a dangerous business in early modern Europe; Savonarola's public execution in Florence in 1498 reminded everyone from common astrologer to exalted natural philoso-

10. For a discussion of the apocalypse as a specific genre and its distinct position in respect to other forms of eschatology, see Christopher Rowland, *The Open Heaven. A Study of Apocalyptic in Judaism and Early Christianity* (London: SPCK, 1982), pp. 9–58.

11. The literature on prophecy in the early modern period is vast, and I have chosen to distinguish it from the general studies of the end of the earth discussed above in note 1. Marjorie Reeves produced a number of important studies, including *The Influence of Prophecy in the Later Middle Ages. A Study in Joachimism* (Oxford: Clarendon Press, 1969); "Some Popular Prophesies from the Fourteenth to the Seventeenth Century," *Studies in Church History* 8 (1971): 107–134; *Joachim of Fiore and the Prophetic Future* (New York: Harper and Row, 1977); and *Prophetic Rome in the High Renaissance Period*, Oxford-Warburg Studies (Oxford: Oxford University Press, 1992). Other studies that emphasize the prophetic and revelatory aspects of apocalyptic thought are: Robert E. Lerner, *The Powers of Prophecy: The Cedars of Lebanon Vision from the Mongol Onslaught to the Dawn of the Enlightenment* (Berkeley: University of California Press, 1983); Donald Weinstein, *Savonarola and Florence: Prophecy and Patriotism in the Renaissance* (Princeton, NJ: Princeton University Press, 1970); Ann Williams, ed., *Prophecy and Millenarianism: Essays in Honour of Marjorie Reeves* (Essex, 1980); Robin Bruce Barnes, *Prophecy and Gnosis: Apocalypticism in the Wake of the Lutheran Reformation* (Stanford, CA: Stanford University Press, 1988); Ottavia Niccoli, *Prophecy and People in Renaissance Italy*, trans. Lydia G. Cochrane (Princeton, NJ: Princeton University Press, 1990).

pher of the risks implicit in making predictions about the future that claimed to rest on divine directives. The interpretation of *signs* of the End of Days, or eschatology, was different from either apocalypticism or prophecy. Most apocalyptic writings in early modern Europe were not actually apocalypses but eschatologies.

Eschatologists like Gabriel Harvey who commented on natural disasters claimed no privileged and direct line of communication from God but instead pointed out the importance of natural signs. Early modern eschatalogical texts attempted to fit recent events into preexisting canons of apocalyptic signs, though some advanced radically new interpretations. Discussing and analyzing signs of the earth's predicted demise was a popular industry in medieval and early modern Europe.[12] The list of pamphlets and books on the new star, on the grand conjunction, and on hosts of natural marvels like monstrous births or rains of blood suggests that anyone might have his views on the subject published. Presses throughout Europe issued one apocalyptic interpretation of contemporary events after another to meet the seemingly insatiable demands of readers. Sheltco à Geveren's *Of the Ende of this Worlde, and the Seconde Commyng of Christ*, for example, was quickly translated by Thomas Rogers from its original Dutch into English and reprinted in London five times between 1577 and 1589.[13] Printing facilitated the formulation of a widely recognized canon of eschatological signs ranging from natural and supernatural

12.   Studies of early modern interest in the eschatology of the final days include: Norman Cohn, *The Pursuit of the Millennium: Revolutionary Millenarians and Mystical Anarchists of the Middle Ages* (New York: Oxford University Press, 1970), which includes information about the later period; Bryan W. Ball, *A Great Expectation: Eschatalogical Thought in English Protestantism to 1660*, Studies in the History of Christian Thought, vol. 12 (Leiden: E. J. Brill, 1975); Richard Bauckham, ed., *Tudor Apocalypse: Sixteenth Century Apocalypticism, Millenarianism, and the English Revolution* (Oxford: Oxford University Press, 1978); Katharine R. Firth, *The Apocalyptic Tradition in Reformation Britain 1530–1645* (Oxford: Oxford University Press, 1979); Heinrich Quistorp, *Calvin's Doctrine of the Last Things*, trans. Harold Knight (London: Lutterworth Press, 1955); William W. Heist, *The Fifteen Signs before Doomsday* (East Lansing: Michigan State College Press, 1952); Robert E. Lerner, "The Black Death and Western European Eschatalogical Mentalities," *American Historical Review* 86 (1981): 533–552; C. A. Patrides and Joseph A. Wittreich, Jr., ed., *The Apocalypse in English Renaissance Thought and Literature* (Ithaca, NY: Cornell University Press, 1984); Werner Verbeke, Daniel Verhelst, and Andries Welkenhuysen, eds., *The Use and Abuse of Eschatology in the Middle Ages* (Louvain: Leuven University Press, 1988).

13.   Tuveson, *Millennium and Utopia*, p. 43.

events to military skirmishes, all of which mounted with alarming speed as the fifteenth century moved into the sixteenth. No longer was knowledge of surprising or unusual events confined to their immediate locales by the limits of word-of-mouth transmission. What this signified in practical terms was that the Catholic church was unable to control the spread of prophetic messages and guide their interpretation. At the Fifth Lateran Council convened in 1512, the Church officially renounced the popular enthusiasm for eschatological speculations that were being printed, circulated, and discussed by those outside its authority.[14]

Just as the printing press was challenging Catholic authorities in the cities of Europe, so the Reformation (whose main proponents made good use of the new printing technology) was establishing alternative forms of worship. The Reformation of the church, coming as it did in the midst of a series of eschatalogical signs, was given a weighty role in Protestant accounts of the world's progress toward redemption. The Catholic church responded with its own program of reform, claiming sole responsibility for leading humanity into the next spiritual age of restitution and rebirth. The jubilee year of 1600 became the focus for Catholic eschatologists. Among the Protestants, the exact time of the Second Coming was more imprecise; many thought that God might begin the next age between 1583 and 1588.[15] This time frame was particularly potent because of its proximity to the great conjunction of 1583, and because prophecies often referred to miraculous events as happening 1,558 years after Christ's birth.[16]

Regiomontanus had been especially influential in raising popular consciousness about the potential import of 1588. Ludovico Guicciardini noted that Regiomontanus "speaketh more at large than the rest" about the eschatological significance of 1588, explaining that "if the wicked worlde shall not perishe, and the sea and Earth bee brought to nothing, at least governementes of kingdomes shall be turned upside downe, and there shall be great lamentation in all places."[17] Elizabeth I launched a

14.　Barnes, *Prophecy and Gnosis*, p. 29.
15.　Studies by Marjorie Reeves, Katharine Firth, and other scholars have explained early modern apocalypticism from a theological and social standpoint. See Reeves, *The Influence of Prophecy in the Later Middle Ages*, Firth, *The Apocalyptic Tradition in Reformation Britain 1530–1645*.
16.　Carroll Camden, "The Wonderful Yeere," in *Studies in Honor of De Witt T. Starnes*, ed. Thomas P. Harrison (Austin: University of Texas Press, 1967), p. 169.
17.　Quoted from the English translation of the dedicatory epistle of Ludovico Guicciardini's *Houres of Recreation, of Afterdinners* (London, 1576) in Camden, *Studies in Honor of De Witt T. Starnes*, p. 169.

halfhearted attempt to curb these eschatological speculations and even issued a formal proclamation against them. Unfortunately, the proclamation was not published until the beginning of the year 1589 and so did little to alleviate the buildup of tensions surrounding the year 1588.[18]

Dee's angelic teachers focused on the year 1588 as the beginning of the eschaton, although they adopted an air of mystery about the precise date after 1588 came and went without serious incident.[19] Dee avidly recorded angelic predictions that the sun would move contrary to its normal course, the brightness of some stars would increase, other stars would fall, and rivers would run with blood.[20] This degree of specificity was fraught with potential disaster, for Dee could be seen as a false prophet or a poor natural philosopher incapable of deciphering eschatological signs if the events did not unfold as predicted. Some of the angels' prophecies proved to be eerily accurate, however. On 5 May 1583 the angels revealed that the beheading of Mary Queen of Scots and a great gathering of ships would soon take place. Four years later Dee looked over his diaries and noted with satisfaction that the foretold events had come to pass: "So she was A° 1587 at Fodringham Castell. And allso the same yere a great preparation of ships against E[n]gland by the King of Spayn the Pope and other princis called Catholik."[21]

From the first recorded angel conversations, Dee was told that "days of tribulations" were afflicting the earth.[22] More detailed information was given by the angelic Governors, who told Dee he was living in the "last tymes, of the *second last world.*" Precisely when these "last tymes" were to conclude, chronologically speaking, was unclear, so Dee clarified this crucial point during a later conversation. When Uriel told him that it was the last of the *"Three last corruptible times,"* Dee hastily interjected with his own understanding of the phrase. Dee understood the angels to mean that the first period "at *Noes flud* ended, the second [began] *at Christ his first* coming and this *is the* third."[23] The angels agreed, and Dee was content.

The angels tended to be more poetic and less specific in their description of the Book of Nature's decay. In an unusually long monologue, the angel Murifri told Dee that:

> The Earth laboreth as sick, yea sick into death.
> The Waters pour forth weepings, and have not moisture sufficient
>   to quench their own sorrows.
> The Aire withereth, for her heat is infected.

18. Ibid., p. 168.   19. See *TFR*, pp. 233, 419–420.   20. Ibid., p. 233.
21. Dee, *AWS* II:389–390.   22. Ibid., pp. 59–60.
23. Ibid., p. 386; Dee's emphasis.

The Fire consumeth and is scalded with his own heat.

The Bodies above are ready to say, We are weary of our courses.

Nature would fain creep again into the bosom of her good and gracious Master.

Darknesse is now heavy and sinketh down together: she hath builded her self, yea (I say) she hath advanced her self into a mighty building, she saith, Have done, for I am ready to receive my burden.

Hell itself is weary of Earth: For why? The son of Darknesse cometh now to challenge his right: and seeing all things prepared and provided, desireth to establish himself a kingdom; saying, "We are not strong enough, Let us now build a kingdom upon earth, and Now establish that which we could not confirm above."

And therefore, behold the end.[24]

When we try to hear these remarks as Dee would have heard them – as a natural philosopher who had spent most of his life trying and failing to understand the natural world – we can see that the impact of Murifri's words must have been considerable. Murifri's references to the "sickness," "infection," "sorrows," and general weariness of the world would have prompted two reactions in Dee: dismay over such ills and hope that the sickness could be cured, the infection driven away, and the sorrows turned into joys. Dee was convinced that he must take steps toward restoring the Book of Nature's condition. The angels explained that their revelations put him on a speedier path toward restitution than others, a path that represented *"new worldes, new peoples, new kings,* & new knowledge of a new Government."[25] To fill in the gaps in the angels' revelations, Dee conversed with his scryer about the decay of nature and its relationship to biblical prophecies. In one of their discussions, Dee and Edward Kelly were talking "of the trubbles and misery foreshewd to be nere at hand."[26]

An interest in the culmination of human and natural history had been a shadowy though persistent feature of Dee's natural philosophy for some time prior to the surviving records of the angel conversations. Two of his published works, the *Monas hieroglyphica* and the "Mathematical Preface," contained references to the decay of the Book of Nature. The "Mathematical Preface," for example, hinted at a link between the deterioration of the earth and God's decision to stop or slow down his process of "Continuall Numbryng." This was a complex idea related to Dee's understanding of the process of creation, which he believed oc-

---

24. *TFR*, p. 4. 25. Dee, *AWS* II:200; Dee's emphasis.
26. Ibid., p. 342.

curred through an act of divine will and speech that continually imbued nature with numerical essences, giving each part a central core that was "Immateriall . . . divine, and aeternall." Since God's numbering of things was also "his Creatyng of all thinges," He could conceivably withdraw a thing's essential "number," decelerate the numbering process, or even stop the process entirely. Then, according to Dee, "that particular thyng shalbe *Discreated*."[27] Dee's treatise contained explanations not only of how things came to be but also of how parts of the natural world could cease to be, providing a natural-philosophic foundation for biblical prophecies about the end of the world. Unfortunately, Dee stopped his explanation there, being reluctant to place too much confidence in the abilities of human beings adrift in "these our drery dayes" to grasp the significance of his arcane ideas.[28]

Dee focused on the deterioration of communication in the *Monas hieroglyphica*, where he mentioned the "deplorable times" facing natural philosophers in the second half of the sixteenth century.[29] Yet there is a tone of cautious optimism in the *Monas hieroglyphica*, his pessimism tempered by his faith that a natural philosopher did not have to sit passively by and watch the world fall apart around him. Dee's hieroglyphic monad was part of his strategy to find a solution to the problems affecting humanity. Claiming that his symbol "reduced or restored" the "quasi-barbaric signs" used by sixteenth-century natural philosophers "to their mystical proportions," Dee set himself up for veneration as that "singular hero" who had "intimately and thoroughly explored the explanations of the celestial events [as well as] the reasons of the rise, the condition, and the decline of other things."[30] The heroic aspect of a natural philosopher's identity was relatively novel in 1564 when Dee wrote the *Monas hieroglyphica*, though it became common in the next century after natural philosophers like Bacon, Galileo, and Boyle had affirmed the progressive nature of scientific knowledge.[31]

Dee made other contributions that he hoped would reverse, or at least arrest, the decay surrounding him. One example is his unpublished proposal in 1582 to reform the calendar, which he referred to as his "Ref-

---

27. Dee, "Mathematical Preface," sig. *jᵛ.   28. Ibid., sig. [tp]ᵛ.
29. Dee, *MH*, p. 117.   30. Ibid., pp. 117 and 121.
31. See Herbert Breger, "*Elias artista* – a Precursor of the Messiah in Natural Science," in *Nineteen Eighty-Four: Science between Utopia and Dystopia*, ed. Everett Mendelsohn and Helga Nowotny, Sociology of the Sciences, vol. 8 (New York: D. Reidel Publishing Company, 1984), pp. 49–72, for an interpretation on the genesis of a progressive role for natural philosophers and scientists.

ormation of tyme."[32] The inaccuracies plaguing the calculation of calendars concerned many scholars of the period.[33] Better observed data, when juxtaposed with accepted early astronomical theories, had resulted in chronological chaos, as the date for Easter crept further and further from the spring equinox and scholars and theologians despaired over their inability to predict the time of Christ's return on such a mutable calendar. Dee's rationale for calendar reform, however, was not founded on a belief that human skills at calculation had improved. Instead, he argued that it was the decline of the cosmos which had placed European astronomers and theologians at a disadvantage. Dee did not rely on events like the 1580 earthquake to justify his position but put forth quite precise and technical astronomical arguments. He noted a change in the declination of the sun since the time of Christ, when "yt was very nere 24 prime mynutes of a Degree more Northerly, and Southerly then now it is." The sun's zenith point had also moved from the time of Christ, as had its distance from the center of the earth.[34] These changes had necessitated a shift in the sphere of the fixed stars, which was forced to respond with "a propre ballancing motion eastward."[35]

Additional information about Dee's beliefs surrounding the mutability of the heavens can be found in his "Mathematical Preface." There he endorsed the practice of astronomy because God had made "the *Sonne, Mone,* and *Sterres,* to be to us, for *Signes,* and knowledge of Seasons, and for Distinctions of Dayes, and yeares."[36] Dee repeated this paraphrased passage from Genesis on the beautifully rendered title page of his plan for calendar reform presented to Elizabeth I: "Let there be LIGHTES in the firmament of the heaven, that they may divide the Day and the Night, and let them be for Signes and Seasons, and for Dayes and yeres."[37] In the "Mathematical Preface," he went on to argue that

32. John Dee, "A Playne Discourse and Humble Advise . . . concerning ye needful reformation of ye vulgar kallendar," Bodleian Library MS Ashmole 1789, ff. 1–40, f. 32a.

33. J. D. North, "The Western Calendar: *intolerabilis, horribilis, et derisibilis;* Four Centuries of Discontent," in *The Universal Frame* (London: The Hambledon Press, 1989), pp. 39–78.

34. Dee, "A playne discourse," f. 7b. Dee claimed to have written a work, now lost, on the distances between celestial bodies and the center of the earth, entitled *De nubium, solis, lunae, ac reliquorum planetarum, immo ipsius stelliferi coeli, ab infimo terrae centro, distantiis, mutuisque intervallis, et eorundum omnium magnitudine* (1551). Dee, CR, p. 26.

35. Dee, "A playne discourse," f. 8a.

36. Dee, "Mathematical Preface," sig. bii^v.

37. Dee, "A playne discourse," f. 2b. The emphasis on "lightes" is Dee's.

mathematics could be used to understand the natural signs mentioned in "Sacred Prophesies," admonishing his readers that it would be wiser to consult the Bible with an astronomical staff as a guide than with the more common reading guide used by schoolchildren.[38] Dee believed that restoring the calendar by bringing observations of the heavens in line with other calculations was one way for a natural philosopher to ensure that astronomical phenomena would correspond to God's revelations in the Bible. Human history, in Dee's view, represented an unfolding of events first prophesied in biblical revelation, and chronological frameworks were vital to predictions of Christ's return to earth.[39]

Dee and the Prophetic Tradition

Despite excellent reasons for the "Reformation of tyme," Elizabeth I and her council rejected Dee's calendar, and Dee grieved that "her Matie will not *reforme it in the best terms of Veritie*."[40] The angels encouraged him to turn from Elizabeth I and her court, and to focus on other potential patrons in Europe who would be more receptive to his reformatory aims. Dee's angel conversations were poised between apocalypse and eschatology, and their delicately balanced position had far-reaching consequences for his reception in the courts of Europe. By 1582 Dee was moving away from the interpretation of eschatological signs and toward a deeper and more dangerous role in the unfolding of the final days: the role of a prophet proclaiming revealed wisdom.

Dee's faith in his angelic protectors gave him the confidence to record and voice their divine revelations, which made him unlike other eschatologists in Protestant countries during the sixteenth century. While many Protestants claimed to speak to God through prayer – a founda-

38. Dee, "Mathematical Preface," sig. bii^v. The text refers to a "festue," a printing error for "fescue," a small stick used by children first learning to read.
39. Scaliger was also deeply engaged in predicting the fulfillment of biblical prophecies, as was Isaac Newton. For Newton's interest in chronology, see Betty Jo Teeter Dobbs, "From the Secrecy of Alchemy to the Openness of Chemistry," in *Solomon's House Revisited: The Organization and Institutionalization of Science*, ed. Tore Frangsmyr (Canton, MA: Science History Publications, 1990), pp. 75–94, especially 83–86. For Scaliger, see Anthony Grafton, *Joseph Scaliger: A Study in the History of Classical Scholarship*, vol. 2: *Historical Chronology*, Oxford-Warburg Studies (Oxford: Oxford University Press, 1994).
40. Dee, *AWS* II:252.

tion of Reformation theology – few received messages in return through multitudes of angels. Intermediaries, both saints and angels, were a feature of Catholic theology, and Protestant theologians tried valiantly, if vainly, to dissuade their congregations from relying on "popish" intercessors like priests or inhabitants of the celestial spheres. In fact, not many Protestants in the sixteenth century claimed to have heard God speak to them in any fashion, direct or otherwise: although claims of direct revelatory experience became widespread in the seventeenth century, Protestants had not yet embraced this possibility in Dee's lifetime.[41] Instead, sixteenth-century Protestants relied on interpreting ancient biblical revelations and prophecies, and deploring the prophecies of contemporary Catholic mystics.

The Catholic church proved equally reluctant to endorse those who claimed to have received direct revelations. Though some mystics were tolerated and even supported, most of these were in holy orders, like Saint Teresa of Avila or Saint John of the Cross, and firmly enmeshed in the institutional hierarchies of the church. Ecclesiastical authorities were experienced in handling wayward and outspoken clerics like Joachim of Fiore and Savonarola: they launched inquisitorial proceedings against them and often preferred to execute them for heresy rather than to support their revelatory claims. Dee was obviously not in holy orders, and his marital state, as the Catholic authorities in Prague pointed out, made him an unlikely candidate for receiving revelations.

The content of the angels' revelations to Dee must have stunned the Catholic authorities in their specificity and unabashed criticism of sixteenth-century life, religion, and government. While other eschatologists talked vaguely about great astrological occurrences and their possible ramifications, Dee claimed to know, with divine certainty, the precise month and year in which miraculous events would occur. Nor did the angels limit themselves to religious matters such as the details of the Second Coming, the end of the world, and the establishment of the new

---

41. A selection of seventeenth-century prophetic and eschatological studies, many focusing on millenarianism, includes Keith Thomas, *Religion and the Decline of Magic* (New York: Charles Scribner's Sons, 1971), pp. 385–434; Ball, *A Great Expectation*, passim; Bernard Capp, *The Fifth Monarchy Men: A Study in Seventeenth-Century English Millenarianism* (Totowa, NJ: Rowman and Littlefield, 1972); Richard Popkin, ed., *Millenarianism and Messianism in English Literature and Thought, 1650–1800*, Clark Library Lectures, 1981–1982 (Leiden and New York: Brill, 1982); James Holstun, *A Rational Millennium: Puritan Utopias of Seventeenth-Century England and America* (Oxford: Oxford University Press, 1987).

Jerusalem. In remarks that would have been treasonable in most countries, the angels shared prophecies about the death of Queen Mary of
Scotland; several invasions soon to face England; the deaths of Philip II,
Rudolf II, and Elizabeth I; and the ascension of Dee's patrons Albert
Laski, Stephen of Poland, and Count Willem Rozmberk to greater positions of power.[42] Political prophecies had the potential to set off popular
revolts, and even the most daring eschatological commentators knew
better than to include clear predictions of their own sovereign's demise –
though predications about the death of rival monarchs were often ignored. Dee should have known better from personal experience than to
mix predictions with matters of state, as he had once been imprisoned
for casting Queen Mary of England's horoscope.[43]

In the final days, Dee's role was to be increasingly prophetic. The
angel Uriel explained: "The Lord hath chosen you to be *Witnesses*,
through his mercy and sufferance . . . in the *offices* and dignities *of the
Prophets* . . . wherein you do *exceed the Temples of the earth*: wherein
you are become separated from the world. . . ."[44] Dee's prophetic role
went hand-in-hand with his increased powers to see and judge the decay
of the Book of Nature. The angels encouraged him to verbalize and
publicize his insights outside the tight circle of the colloquia, and scolded
him when they thought he should have made "a more ample declara-
tion" of their remarks.[45] The angels dictated that Kelly, though also
touched with the powers of prophecy, should limit himself to seeing
visions in the showstone rather than sharing his visions with a wider
audience – a wise course given Kelly's volatility. While the angels promised to make Kelly "a great Seer: Such an one as shall judge the circle of
things in nature," they also told him that "heavenly understanding, and
spiritual knowledge shall be sealed up from thee in this world."[46]

Dee embarked upon his potentially dangerous prophetic activity because he believed himself heir to the prophetic biblical tradition of Enoch, Elias, Esdras, and John the Divine. The mention of these prophets
in Dee's angel conversations gave his revelations a greater sense of authority and the prophecies concerning the final days a greater urgency.
Each prophet enjoyed privileged communication with God or angels that
resulted in a certain knowledge of the world as it was, or the world as it
was to be after God's last judgment. Each prophet had served as a

42. See Dee, *AWS* II:389–390; *TFR*, pp. 380–381, 398, 419–420, *12.
43. Peter French, *John Dee: The World of an Elizabethan Magus* (New York:
    Routledge and Kegan Paul, 1972), p. 34.
44. *TFR*, p. 233.   45. Ibid., p. 400.   46. Ibid., p. 61.

"divine scribe," obediently recording the information received from God and the angels in books of wisdom. Though Esdras and John were important models for Dee, he found Elias and Enoch more fascinating. His interest in these two mysterious prophets was shared by other contemporary natural philosophers, including Guillaume Postel and Paracelsus.[47]

In the Judeo-Christian tradition, Enoch's important position was based on a few lines in Genesis and Ecclesiasticus that related how Enoch "walked with God" and then "was no more, for God took him" (Genesis 5:21–24; Ecclesiasticus 44:16). Texts attributed to Enoch circulated in the early Christian centuries before disappearing.[48] Even with the loss of these precious texts, Christian commentators were quick to expand upon the references to Enoch, thereby bolstering his profile. The angels shared their knowledge of Enoch with Dee, information which happily corresponded to early modern perceptions of the prophet. To Dee, the greatest knowledge that Enoch possessed was his special, post-lapsarian ability to read the Book of Nature. Enoch's eyes, according to the angels, were opened by the Lord so "that he *might see and judge the earth,* which was unknown unto his Parents, by reason of their fall." In addition, Enoch was shown "the *use of the earth*" and was "full of the spirit of wisdom."[49] As Dee longed for mastery over the Book of Nature, it is not surprising that he was eager to discuss Enoch with the angels. Enoch's mysterious departure from the earth was equally intriguing. Most of Dee's contemporaries believed that God took the prophet to Paradise or heaven, where he lived free from sin in full knowledge of God. Since such communion with divinity was thought to occur only after death, Enoch's acceptance into God's grace was even more extraordinary.

According to scripture, Enoch was joined in Paradise by another figure prominent in the angel conversations: the prophet Elias (also known as Elijah), whose story was told in 1 Kings 17–19. Elias, like Enoch, was transported to heaven prior to death. He was especially revered in the Jewish tradition, because references in Malachi 4:5–6 suggested that

47. Marion Leathers Kuntz, *Guillaume Postel, Prophet of the Restitution of All Things: His Life and Thought* (The Hague: Martinus Nijhoff Publishers, 1981), pp. 78–91; Breger, *"Elias artista,"* passim.

48. For a history of the Enochian texts, their disappearance, and discovery, see R. H. Charles, "The Book of Enoch," in *Apocrypha and Pseudepigrapha of the Old Testament* (Oxford: Oxford University Press, 1913), 2:163–281.

49. TFR, p. 174.

Elias would return from heaven before the Final Judgment to bring the Israelites to repentance.[50] Elias's association with the final days also meant that his return was one of the eagerly awaited signs that the eschaton was near. His imminent return was confirmed by the angels, who told Dee that soon the "Prophets of the Lord shall descend *from* Heaven." Dee connected this remark to Elias immediately, noting in the margins of his angel diaries that the prophets were "in Paradise, they were carried upward, especially Elias."[51]

Elias was an important figure in works concerned with the occult sciences as well as early modern eschatologies. Paracelsus (1493?–1541), the alchemical physician whose works appeared frequently in Dee's library, referred to Elias in his writings. Paracelsus drew on biblical ideas and commentaries to create *Elias artista*, who would return to the world and reveal occult aspects of the Book of Nature. Paracelsus linked the figure of *Elias artista* to his belief in a utopian future for the world that would restore both the Book of Nature and humanity.[52] Before utopia could be achieved, however, *Elias artista* must return so that, through him, God could "reveal much that is strange, much that has not yet been uncovered and revealed to us, things of which we are all ignorant."[53] Paracelsus's fascination with *Elias artista*'s secret knowledge helped to ensure that alchemists and occultists kept the tradition of Elias alive.[54]

Guillaume Postel (1510?–1581) was also fascinated by the prophet; indeed, evidence suggests that he believed that he himself was *Elias artista*.[55] A female mystic known as the "Virgin of Venice" first associated Postel with Elias between 1546 and 1547, when she described a number of changes about to affect the earth, including the formation of a universal church and state, and humanity's return to a life of perfection.[56] This rebirth would begin with one "who lives and acts in the

---

50. For more on Elias, see John Van Seters's article "Elijah," in *The Encylopedia of Religion*, ed. Mircea Eliade (New York: Macmillan Publishing Company, 1987), 5: 91–93, for a summary and bibliography, and Breger, "*Elias artista*," passim.
51. TFR, p. 61.   52.   See Breger, "*Elias artista*," pp. 54–55.
53. Paracelsus, *Sämtliche Werke*, 3:46; translated and discussed in Breger, "*Elias artista*," p. 54.
54. Breger, "*Elias artista*," pp. 56–57.
55. Kuntz, *Guillaume Postel*, p. 175.
56. For more information on the Virgin of Venice, see Marion Leathers Kuntz, "The Virgin of Venice and Concepts of the Millennium in Venice," in *The Politics of Gender in Early Modern Europe*, ed. Jean R. Brink, Allison P. Coudert, and Maryanne C. Horowitz, Sixteenth Century Essays and Stud-

spirit of Elias," who she disclosed was Guillaume Postel.[57] Postel's role in the restitution of nature and the world as revealed by the Virgin of Venice was nearly identical to the role the angels assigned to Dee.

In addition to the restitution of nature, Dee was expected to restore religious unity. Popular hysteria about the coming woes escalated as the religious divisions of Europe deepened, and most of Dee's contemporaries linked the two crises. English chroniclers John Stowe and William Camden both mentioned the desperate tenor of the times, religious discord, and the hoped-for reconciliation between Catholics and Protestants. "All Europe stood at gaze, vehemently expecting more strange and terrible alterations," wrote John Stowe, "than ever happened since the world began."[58] Protestant and Catholics alike looked forward to the day when their own religion would triumph and become the universal faith.

Where Dee's angel conversations and his role as prophet fit into this puzzle of eschatological and reformist ideas has always been difficult to fathom. Fully embracing Yates's belief in Hermeticism as the dominant intellectual framework within which any understanding of Dee must take place, French suggested that Dee believed in a "Hermetic religion of love" that would heal the divisions between Protestants and Catholics. Clulee described Dee's "rather flexible religious convictions" as a combination of a "Protestant view of salvation" with a "latitudinarian . . . acceptance of varying forms of religious observance." Both authors, however, also noted Dee's personal interest in reforming and restoring religion to its "true" sense and purpose.[59] Dee did not limit his audience to Protestants or Catholics, but believed that the prophecies were "to be (by us) published . . . all the World over."[60] The switches Dee made from Protestantism to Catholicism and back to Protestantism again have raised provocative questions about his "true" beliefs.

Instead of choosing between Catholic and Protestant faiths the angels told Dee that an alternative course was open to him: the establishment of a new, angelically revealed universal religion that would accompany his universally applicable natural philosophy.[61] Dee was not willing,

---

ies, vol. 12 (Kirksville, MO: Sixteenth Century Journal Publishers, 1989), pp. 111–130 passim, especially 123.

57. Quoted in Kuntz, *Guillaume Postel*, p. 78.

58. John Stowe, *Annals of England* (London, 1632), p. 743; discussed in Camden, "The Wonderful Yeere," p. 172.

59. French, *John Dee*, pp. 123–125; Clulee, pp. 193 and 208.

60. Dee, *AWS*: 376–377.

61. See, for example, *TFR*, p. 59; Dee in Josten, p. 245.

therefore, to subject himself to the authority of the Catholic church, but he was not an advocate of the Protestant faith either. Perhaps his reluctance to adhere too closely to the new religion stemmed from Uriel's dire prophecies concerning Luther, Calvin, and "all that have erred, and wilfully runne astray" from the Catholic church, which was "by Peter brought to Rome, [and] by him, there taught by his Successors." All those who "erred" from the Catholic church were destined to be sent into the "Hell fire" by Christ's final judgment.[62] Despite these remarks, the angels did not endorse Catholicism completely and made a special point of berating the clergy "that inhabit the holy City, and usurp the authority of the Highest."[63]

Dee believed that all postlapsarian religious beliefs were incapable of restoring the church to what God had originally intended. His religious reformation, on the other hand, promised a return to a truly universal faith that would include the Jews and Muslims as well as Catholics and Protestants. Dee had been interested in the question of religious unity for some time. He joined the fray of early modern conversion debates with a book, now lost, entitled *De modo Evangelii Jesu Christi publicandi, propagandi, stabiliendi inter Infideles Atlanticos* (1581), a title which suggests a Native American orientation. Dee's inability to find a spiritual haven for himself in either the Catholic or the Protestant camp is clear in his plans for the distribution of the treatise. According to the *Compendius Rehearsal*, the work was first presented to Queen Elizabeth, then to other key figures in her government – a good strategy for a natural philosopher seeking a position within an officially Protestant country. The other recipients were the two preeminent Catholic powers in Europe: Philip II of Spain and the pope.[64] The angel diaries suggest that Dee did give the document to the king of Spain, as on 17 November 1582, the year after *De modo Evangelii Jesu Christi* was completed, he spoke with the angels about the "prayse and advancem^t of Gods Glorie, with Philip the Spanish King."[65]

More crucial to most early modern Europeans was the conversion of the Jews. Many of Dee's remarks about conversion in the angel conversations concerned them and combined a paradoxical though fairly common early modern blend of anti-Semitism with an intense interest in secret, mystical Hebrew knowledge. This combination was especially apparent when it came to the "lost books" that Dee believed that Jews possessed, especially those of Esdras. He earnestly asked the angel II

62. *TFR*, pp. 411–412.   63. Ibid., p. 233.   64. Dee, *CR*, p. 26.
65. Dee, *AWS* II:174, and Whitby's remarks on p. 391, which also link these remarks to the lost work of 1581.

how he could "trust anything in the Jues hands," since they were "a stiffnecked people and dispersed the world over."[66] The angel replied with a vague prophecy about the number eighty-eight, which Dee interpreted as an indication that the year 1588 might bring about the Jews' anticipated conversion.[67] Despite these remarks, the angels took recourse in positive examples set by Old Testament Jews when Dee or his scryer misbehaved. On one occasion Edward Kelly asked if he could have the authority of the angels' revelations verified by his Jesuit confessor. An angel replied tartly that they possessed greater credibility than a "fleshly Priest." In addition, the angel reported: "If the Angels that have appeared unto you [Edward Kelly], had appeared also unto the Jews, saying, Crucifie not the Son of God, they would not have done it. For though they believed not man, yet would they have believed an angel."[68] Dee duly noted the rebuke in the margins of the angel diaries.

Dee continued to search for a sponsor for his conversion strategies throughout the period of the angel conversations. Neither Elizabeth nor Philip expressed any interest in his plans for religious conversion, but Dee was more successful in convincing Count Laski, who actually took part in the conversations, to accept his plans to unify the world under a single faith. The angel Jublandace prophesied that Laski would achieve "great victories, in the name, and for the name of his God," and that under Laski's leadership the "*Jews . . . shall taste* of this Crosse: And with this *Crosse shall he* [Laski] *overcome the Sarazens,* and *Paynims.*"[69] As we have seen, Laski fell from God's favor, and though both Rudolph II and Steven of Poland were invited to spearhead the angelic movement toward a single faith, both declined to assume an active role in Dee's reformed religion.

Despite these setbacks, Dee's interest in a unified faith remained unshaken, and he continued to believe that the angelic revelations provided the basis of a holy doctrine which could "restore, and . . . put in good order, the people and flock of God."[70] The tenets of the angelically revealed, "true" reformed religion represent a fascinating combination of Catholic and Protestant doctrines that promised to reconcile all religious factions of the early modern period. As an angelic voice told Dee

---

66. Ibid., p. 334.   67. Ibid., p. 336.   68. *TFR*, p. 386.
69. Ibid., p. 22; Dee's emphasis.
70. Dee in Josten, p. 244. I am indebted to Samantha Meigs for discussing the implications of Dee's complicated strategy for religious reform, and for generously sharing her knowledge about early modern popular religion and the Catholic and Protestant theological debates that took place during the period.

and his associates while they were in Prague: "Whosoever wishes to be wise may look neither to the right nor to the left; neither towards this man who is called a catholic [*sic*] nor towards that one who is called a heretic . . . but may he look up to the God of heaven and earth and to his son, Jesus Christ. . . ."[71]

What this religious compromise actually meant, of course, was that Dee's new religion pleased no one. The angelic doctrine endorsed the power of prayer, for example – a dominant feature of Protestant theology – while reporting the dire consequences that had befallen Luther, Calvin, "and all that have erred, and willfully runne astray" from true Catholic church.[72] While the Roman Catholic church deserved Dee's loyalty, the pope did not; because the authority of the angels far outweighed that of the pope, Dee was told to consult only with the angels for spiritual advice and counsel.[73]

Dee's special status gave him the capacity to monitor and gauge the progress of humanity toward restitution, and this represented one of the most unorthodox and unique aspects of the new, reformed theology.[74] The angels also revealed their understanding of election, free will, and predestination, blending Catholic and Protestant positions into an artful compromise.[75] They confirmed repeatedly that Dee was one of the "elect," yet his election could be affected by his conduct. As the angels told Dee, "the elect and chosen may erre and go astray."[76] Man, through free will, "draweth out of order, turneth from the mark, refuseth that which is good, and through the burden of his flesh, inclineth unto evil."[77] The keys to salvation for the elect were perseverance and prayer, rather than confession and intercession.

One of the thorniest debates between Catholics and Protestants involved the celebration of the Eucharist, a crucial question the angels did not neglect. Dee and Kelly actually partook of the Catholic mass while they were in Prague.[78] But before taking the sacrament in the Catholic church, Dee entered into a theological discussion with the angels about transubstantiation and gained a fresh understanding of the Eucharistic celebration and its role in the true religion.[79] Dee's desire to seek angelic spiritual counsel underscores the ways in which his religious views and practices were being shaped by the angel conversations. His angelic advisers demanded that he "lay . . . religion aside" and "simply and

---

71. Dee, quoted in Josten, p. 245.   72. *TFR*, p. 411.
73. Ibid., p. 386.   74. Dee, *AWS* II:378.   75. *TFR*, p. 383.
76. Ibid., p. 384.   77. Ibid.
78. Ibid., pp. 397, 399, 401; French, *John Dee*, p. 121; Clulee, p. 224.
79. *TFR*, pp. 371–373.

nakedly follow the steps of true *Faith*," as set out in the examples of the Apostles.[80] Dee was not at all clear what this entailed, and questioned the angels closely on the symbolism and significance of the Eucharist as Christ originally imparted it to His disciples. From Dee's questions, it is apparent that the Catholic authorities in Prague had demanded that the natural philosopher and his scryer acknowledge transubstantiation before taking the sacrament. The angels, once again, crafted an admirable middle course between Catholic and Protestant views on the subject that enabled Dee to take part in the Catholic ceremony while maintaining his role as the prophet of a new, universal religion that would also be acceptable to Protestants.

The angels told Dee that Christ had ministered "his true body" to his disciples, so that "the *Minister*, using the office and person of Christ . . . pronouncing the words, doth also give unto the people not Bread, *but the true body*."[81] Up to this point, the angelic doctrine was decidedly Catholic, especially in its emphasis on the power of the priest's words to cause a change in substance from simple bread to the body of Christ.[82] The Catholicism of the angelic doctrine was reinforced when Dee was told that the "body" of Christ was present in both the transubstantiated bread (which had become Christ's "flesh") and in the transubstantiated wine (which had become Christ's "blood"). Dee was relieved that Christ was fully present in both the bread and the wine, because the Catholic rite usually provided only bread to the laity. The angels assured him that "it is no offence to God, to receive . . . one [substance] . . . onely."[83] While these aspects of the angelic doctrine coincided with prevailing Catholic practices, the angels also placed a Protestant emphasis on the importance of each individual's faith in the act of receiving the Eucharist. Though irrelevant to Catholicism, in which the bread became a sacrament because of the actions of the priest, the angels told Dee that the Eucharist was sacramental only if the individual *believed* it was sacramental. Thus, Dee was free to participate in the Catholic mass because "by faith" he believed it was "the true body" of Christ.[84]

This carefully charted and almost universally heretical angelic religion was difficult to adopt and practice, and, over time, Kelly became a vocal opponent of the angels' plans for the new faith. As he did every aspect of the angel conversations, Dee recorded the specifics of Kelly's displea-

---

80. Ibid., p. 372.   81. Ibid.
82. Horton Davies, *Worship and Theology in England*, vol. 1: *From Cranmer to Hooker 1534–1603* (Princeton, NJ: Princeton University Press, 1970), p. 163.
83. *TFR*, p. 372.   84. Ibid., p. 373.

sure, despite Dee's disagreement with his scryer's extreme outbursts regarding the "manifold horrible Doctrine." It is impossible to say why Dee's scryer felt obliged to make such comprehensive and damning statements, unless it was to clear himself in the event the apostolic nuncios did manage to take the pair into custody. Whether for self-preservation or self-vindication, Kelly condemned the angels' efforts to persuade him that various heretical ideas were true, including that "Jesus was not God," that "no prayer ought to be made to Jesus," that a "mans soul doth go from one body to another childes quickening or animation," that there was no Holy Ghost, and that "the generation of mankind from *Adam* and *Eve*, is not an History, but a writing which hath an other sense."[85] When and where the angels discussed these aspects of their doctrine with Kelly is not clear, for they do not appear anywhere in the extant angel diaries. With its Arian and Pelagian ideas, the angelic doctrine, as represented by Kelly, would indeed have been dangerous. But to Dee's great delight, Kelly was soon "converted" to the angelic doctrine and religion. Dee went on to address pointed prayers to Jesus and the Holy Ghost apologizing for Kelly's behavior, thereby divorcing himself through prayer from most of what his scryer had said.[86]

Dee was neither a Catholic nor a Protestant, but was following his own religious convictions, which he believed might heal the deep divisions of Western Christianity and convert nonbelievers to the true faith. Many scholars have suggested that his religious vacillation between Catholicism and Protestantism was influenced by Familism. Any definitive association between Dee and the Familists will probably remain conjectural, but there are striking similarities between his angelic religion and Familism with its emphasis on revelation, universality, and communication. Hans Niclaes (1501–1581) founded Familism around the year 1540 after receiving a revelation from God calling him to found a new religion.[87] Like the doctrine outlined to Dee by the angels, Familism was a revealed theology that hoped to unite people of all faiths within a single, reformed religion. Also, it attempted to recall the prelapsarian relationship between humanity and divinity.[88] More specifically, the religion believed in the importance of faith over reason, the verity of revelation, and the ability of all people to become prophets because of their religious

---

85. Ibid., p. 164.   86. Ibid., p. 165.
87. B. Rekers, *Benito Arias Montano (1527–1598)* (London: Warburg Institute, 1972), p. 73.
88. Alastair Hamilton, *The Family of Love* (Cambridge: James Clarke and Company, 1981), p. 34.

experiences. Because Familists believed that religious and civil harmony should exist, they were allowed to adopt the official state religion of any country, as long as they adhered to the beliefs of Familism in their hearts. Thus, practicing Familists were not required to exhibit outward signs of affiliation to their church.

Most scholars believe that the Familist doctrines entered England from the Low Countries at the accession of Mary I in 1553 – the same year that Dee returned from Louvain. Dee's associates in Louvain and Paris – Ortelius, Mercator, Postel, and Frisius – were either Familists or knew of their teachings, so it is likely that he was introduced to their beliefs while in Europe in the 1540s and 1550s. This was not the only way in which Familist ideas might have filtered into Dee's angel conversations, however. The center of European Familism was Emden, a small town in East Frisia where Niclaes settled and began to communicate his revelations to a wider audience.[89] The other religious leader in the town of Emden during this time was a Protestant reformer and friend of Erasmus, Johannes à Lasco. The reformer of Emden was related to Count Laski, a key member of Dee's colloquium of angels, and the association between the count and his reforming uncle might have helped to consolidate Laski's prominent position in the angels' revelations concerning the new religion.[90]

Despite the aura of secrecy which surrounded them, a knowledge of the more sensational Familist tenets began to seep into popular consciousness through Dutch scholars and booksellers in European cities like London. While Dee might have been well acquainted with the subtleties of Familist ideas about universal religion as a result of his time spent in Louvain, for example, Edward Kelly was probably more familiar with the popular notion that Familists practiced communal living and even shared their wives rather than adhering to the monogamous bonds of holy matrimony. This charge was always strongly refuted by the Familists themselves, who pointed out that their community was one of "spirit," not property and body. Nonetheless, discussions of their lifestyle appear in many polemics, especially those published in England.[91] One of the most bizarre and infamous aspects of Dee's angelic doctrine

---

89. Ibid., p. 32. Laski and Dee visited Emden on their way from England to Cracow and remained in the town from 17 October 1583 to 20 October 1583.
90. Hamilton, *Family of Love*, p. 32.
91. Leon Voet, *The Golden Compasses:* The History of the House of Plantin moretus (New York: Abner Schram, 1969), 1:25–26.

might relate to this suspected Familist practice: namely, the angels' instruction that Dee and Kelly should share all things between them, including their wives.[92] This was one of the angels' most difficult requests, and it stretched Dee's faith in the angel conversations to the breaking point.

Dee did not adhere to any of the many religions present in Europe during the sixteenth century. In his eyes, no religious doctrine was infallible except for that espoused by the angels, which promised to restore religion to a central doctrine revealed by God and capable of unifying many faiths into a single faith. This angelic religion was the religion of the world to come, the newly restored and perfected natural world that would follow the End of Days. Dee's own role in the perfection of nature, however, would depend upon his mastery of the divine language of creation and a new exegetical tool, "the true cabala of nature."

92. *TFR*, pp. *9–*22; Deborah E. Harkness, "Managing an Experimental Household: The Dees of Mortlake," *Isis* 88 (1997): 257.

5

"The True Cabala"

Reading the Book of Nature

Of this fair volume which we "world" do name,
If we the sheets and leaves could turn with care,
Of Him who it corrects, and did it frame,
We clear might read the art and wisdom rare?
Find out his power which wildest powers doth
    tame,
        His providence extending everywhere
. . . .
But silly we (like foolish children) rest
Well pleased with colored vellum, leaves of gold,
Fair dangling ribbons, leaving what is best,
On the great Writer's sense ne'er taking hold;
    Or if by chance our minds do muse on aught,
      It is some picture on the margin wrought.
      —William Drummond of Hawthornden
           (1585–1649)

To have wisdom is a gift from God: To have
knowledge comes from the creation and created
things.
      —John Dee's angel conversations

John Dee had crucial responsibilities in the final days and the restitution of all things as an interpreter of both the angels' revelations and the Book of Nature, and as a prophet spreading news of the upcoming changes to the elect. Yet the moral fall of humanity and the decay of the natural world had rendered the Book of Nature illegible and beyond comprehension. As Nalvage told Dee, "all things grew contrary to their creation and nature," which made God withdraw from the world, because "the Elements are defiled, the sons of men [are] wicked, their bodies [have] become dunghills."[1] To help Dee draw closer to God and

1. *TFR*, pp. 76–77.

nature, the angel Raphael gave him a new exegetical tool: the true cabala of nature, described as "simple, full of strength, and the power of the holy Ghost." Using it, Dee would be able to decipher the rapidly disintegrating Book of Nature and accurately interpret the eschatological signs embedded there.

The cabala of nature enabled Dee to have both wisdom (a revealed gift) and knowledge (an acquired understanding of the complexities of the created world). This mixture of wisdom and knowledge, of revelation and natural philosophy, lies at the heart of the cabala of nature. With his admission into the colloquium of angels and his instruction in the cabala of nature, Dee believed that he had achieved a state of intellectual virtuosity similar to that enjoyed by Adam "before his fall," who "not onely knew *all* things, but *the measure and true use thereof.*"[2] As Dee continued to employ the cabala of nature, he believed the world was moving toward the long-awaited eschaton that would heal and restore it. Dee's mastery of the cabala of nature was based, first and foremost, on his skill with the divine language that Adam and God had shared in Paradise. The divine language then enabled Dee to reform and restructure the existing Hebrew and Christian cabalas, restoring this science to its true power and intent.

Reclaiming the Divine Language

Dee's angel diaries are filled with a preponderance of mystical letters, numbers, and strange words. They leap from the pages, their strange squiggles and incomprehensible phrases catching the eye as surely as Dee's showstone caught the rays of light in his study. These passages contain what Dee believed was the divine language of the Book of Nature lost to humans since the days of Adam. Dee referred to the language as the "celestial speech," the "language of the angels," the "speech of angels," the "speech of God," and the "language of Adam."[3] Dee's use of these terms indicates his understanding that the divine language preceded Adam's expulsion from Paradise, thus escaping the imperfections that plagued the world after the Fall.[4] Modern scholars,

2. Dee, *AWS* II:226–227; Dee's emphasis.
3. Ibid., pp. 232, 332; *TFR*, p. 19.
4. Modern scholars have complicated the situation by using the term "Enochian" to denote the language of the angel conversations, thereby referring to Dee's interest in the biblical figure Enoch as well as to a particular revealed book which Dee received during the angel conversations. Dee uses Enochian to describe a tablet or book reputedly given to Enoch, but seldom

far removed from Dee's worldview and hampered by the scattered and imperfect nature of the manuscripts, have not adequately examined this aspect of the angel conversations, though Nicolas Clulee has suggested that Dee's fascination with the angels was based on his intention to recover Adam's divine language and ability to communicate with the heavens, and that his interest was part of a widespread fascination with recovering God's language.[5]

Early modern humanists, intrigued by the possibility of restoring ancient texts, explored the history of God's creative language and its subsequent corruption. The first biblical reference to language appears in the initial passages of Genesis, where God created the world through speech. Early modern authorities agreed that God shared His creative language with the angels and Adam in Paradise. After Adam sinned, his mastery of the divine language was lost along with his ability to communicate with the Book of Nature and God. Thus the disintegration of Adam's linguistic and communication skills was inextricably tied to the fall of mankind and the decay of the natural world. The precise moment of Adam's "linguistic Fall" differed, however, from author to author. Some believed that Adam left Paradise still possessing divine language, which gradually deteriorated over time as humanity moved farther and farther from God's intended plan. Others adhered to a more cataclysmic mechanism to explain humanity's loss of the divine language, suggesting that either Noah's flood or the destruction of the Tower of Babel had erased God's language from the earth.[6] The Tower of Babel was also held responsible for the confusion of earthly tongues and the division of the earth's peoples into nations with linguistic differences.[7] What had once been a united people with a single language became seventy distinct

uses the word outside of this direct context. See Clulee, 221; Donald C. Laycock, *The Complete Enochian Dictionary* (New York: S. Weiser, 1978), passim.

5. Clulee, pp. 203–230, esp. 213.
6. For a discussion of these ideas and their connections to other cultural and intellectual concepts, see Allison Coudert, "Some Theories of a Natural Language from the Renaissance to the Seventeenth Century," in *Magia Naturalis und die Entstehung der Modernen Naturwissenschaften* (Wiesbaden: Franz Steiner Verlag, GMBH, 1978), pp. 56–114 passim. Arnold Williams, *The Common Expositor: An Account of the Commentaries on Genesis, 1527–1623* (Chapel Hill: University of North Carolina Press, 1948), surveys views held by Dee's contemporaries.
7. Genesis II:1–9. For the role of these passages in later thought, see Robert Graves and Raphael Patai, *Hebrew Myths: The Book of Genesis* (New York: McGraw-Hill, 1963), pp. 125–129.

nations with seventy separate tongues. The language Adam spoke after the Fall, already weakened by the Flood, was further dissipated.

Commentators on Genesis also disagreed about which extant early modern language was most closely related to the divine language. Most of the scholars Dee consulted, including Johannes Reuchlin, thought that the likeliest candidates were the three biblical languages (Hebrew, Greek, and Latin), with Hebrew at the forefront.[8] Guillaume Postel was committed to Hebrew as the source for all terrestrial, postlapsarian languages and argued that it was closest to the divine language, both etymologically and chronologically.[9] Agrippa agreed with Postel, calling the Hebrew alphabet "the foundation of the world," and explaining in his *De occulta philosophia* how every creature and language in the world was related to the divine language through Hebrew.[10] More creative answers followed: Swedish natural philosopher Andreas Kempe (1622–1689) argued that God spoke to Adam in Swedish, Adam spoke to the animals in Danish, and the Serpent spoke to Eve in French.[11]

In the angels' conversations with Dee a compromise was reached among divergent theories about the survival of the divine language. The angels revealed that Adam knew two languages, and that both had been lost. The first was a "greater" language that Adam used to communicate his deteriorating "knowledge of God [and] his Creatures" to his descendants; the second was a "lesser" language that Adam used in everyday speech and communication. The first language was formed from remnants of the divine language shared by God, the angels, Adam, and the Book of Nature. With his "greater language" Adam partook of the spirit of God, knew all things in the created world, spoke of them properly, named them correctly, and conversed with God and the angels.[12] The angels were forthcoming about the precise nature of Adam's linguistic Fall and his use of the second, lesser language, reporting that Adam lost most of his mastery and understanding of the "greater" language after disobeying God's commands and eating fruit from the Tree of Knowl-

---

8. From Reuchlin's correspondence, quoted in François Secret, *Les Kabbalistes chrétiens de la Renaissance* (Paris: Dunod, 1964), p. 47.

9. Kuntz, Marion Leathers, *Guillaume Postel, Prophet of the Restitution of All Things: His Life and Thought* (The Hague: Martinus Nijhoff Publishers, 1981), p. 34.

10. Agrippa, *DOP* I:162

11. See Claes-Christian Elert, "Andreas Kempe (1622–1689) and the Languages Spoken in Paradise," *Historiographica Linguistica* 5 (1978): 221–226.

12. *TFR*, p. 92.

edge. During his first days after his expulsion from Paradise, Adam retained a sense of the divine language and began to craft an ancient form of Hebrew from it – which declined along with the rest of the world during the intervening centuries.[13] Early modern Hebrew was merely a descendant of this second Adamic language, and thus retained only the barest traces of the divine language.

Dee, a bibliophile to the last, asked the angels whether Adam had written anything in his postlapsarian language, alluding perhaps to biblical passages in Jude 14–15, where the books of Adam and Enoch were mentioned.[14] The angels' reply was ambiguous. One angel said that all traces of Adam had been lost in Noah's flood, along with all remnants of the first, prelapsarian language. "Before the flud," the angel told Dee, "the spirit of God was not utterly obscured in man. Theyr memories were greater, theyr understanding more clere, and theyr traditions, most, unsearchable."[15] What prompted Dee to ask the question was his belief that he possessed Adam's *Aldaria*, or *Book of Soyga*. Uriel confirmed that the book was "revealed to Adam in paradise by God's good angels,"[16] and the angels' later comments help to clarify why Dee believed the book was so valuable – though they contradict the angels' other explanations of the moral Fall and its linguistic consequences. First, the angels told Dee that the *Book of Soyga* contained the language Adam was taught "by infusion" in Paradise; the contents thus predated the Fall and expulsion. Unlike the divine language, however, the language contained in the *Book of Soyga* was used after the Fall, and it continued to be used until the destruction of the "Ayrie Tower" – the Tower of Babel.[17]

Adam was the key figure in the angels' discussions of the divine language because he had used the language to name all the creatures in the garden, bestowing true and essential names on all animals and plants.[18] After Adam's fall, only select individuals were made aware of the divine language, and only fragments of it were revealed. The angels

---

13. Ibid., p. 92.   14. Dee, *AWS* II:333.   15. Ibid.   16. Ibid., p. 17.
17. Ibid., p. 332.
18. This important idea stems from biblical tradition and figures prominently in Neoplatonism and the cabala. For a modern study that explains the genesis of the idea and its inclusion in early modern natural philosophical, magical, and cabalistic texts, see Allison P. Coudert, "Forgotten Ways of Knowing: The Kabbalah, Language, and Science in the Seventeenth Century," in *The Shapes of Knowledge from the Renaissance to the Enlightenment*, ed. D. R. Kelley and R. H. Popkin (Dordrecht: Kluwer Academic Publishers, 1991), pp. 83–99, especially 91–93.

told Dee he was receiving an unprecedented amount of the divine language, a fact that provoked an enormous amount of excitement. Not even the powerful biblical prophets Esdras and Enoch had known the language in its entirety while they were on earth, yet Dee was receiving the language's alphabet along with texts written in the language, free from the vicissitudes of earthly corruption and decay.

Dee's attention to Adam's use of the divine language related to his interest in the esoteric Jewish wisdom of the cabala. Because scholars previously believed that Dee's knowledge of Hebrew and the cabala was slight, few comparisons have been made between the angel conversations and Jewish or Christian cabalistic systems and ideas.[19] While Clulee outlined the connections between Dee's earlier intellectual interests and his later participation in the angel conversations, he deemphasized the potential place of cabalistic ideas in the angel conversations, thus restricting possible connections between them and natural philosophy.[20] Though Clulee's purpose was not to analyze Dee's interest in the angels comprehensively, his brief discussion of the apocalyptic and redemptive nature of the angel conversations points to several fascinating associations between Dee's philosophy of nature and his attempts to communicate with the angels.

Dee related his efforts to learn God's divine language to his study of the cabala and Hebrew as early as 20 March 1582. In marginal remarks, he referred easily to passages in several authoritative cabalistic and Hebrew texts, including works by Johannes Reuchlin and Petrus Galatinus.[21] In a later exchange concerned with the secret names of God, a popular topic in works of Christian and Jewish cabala, Dee told the angels that he had "re[a]d in the Cabala of the Name of God . . . [represented in] 42 letters" but had not come across any reference to the name of God represented in only forty letters, which the angels were proposing.[22] To ensure that the angels were clear about this discrepancy, he inscribed all of God's names using forty-two Hebrew characters (Figure 5) in his diary. The inclusion of the names of God in the angel diaries suggests that the diaries may contain additional information to help modern scholars assess Dee's familiarity with Hebrew; other passages in Hebrew, some quite lengthy and others consisting of only a word or phrase, may also be of use.[23] When this information from the angel

---

19. Clulee, p. 214.   20.   Ibid., pp. 215–216.
21. Dee, *AWS* II:65. For more information on Galatinus, see G. Lloyd Jones, *The Discovery of Hebrew in Tudor England: A Third Language* (Manchester: Manchester University Press, 1983), pp. 133–134.
22. Dee, *AWS* II:65.   23.   For example, ibid., pp. 7 and 65.

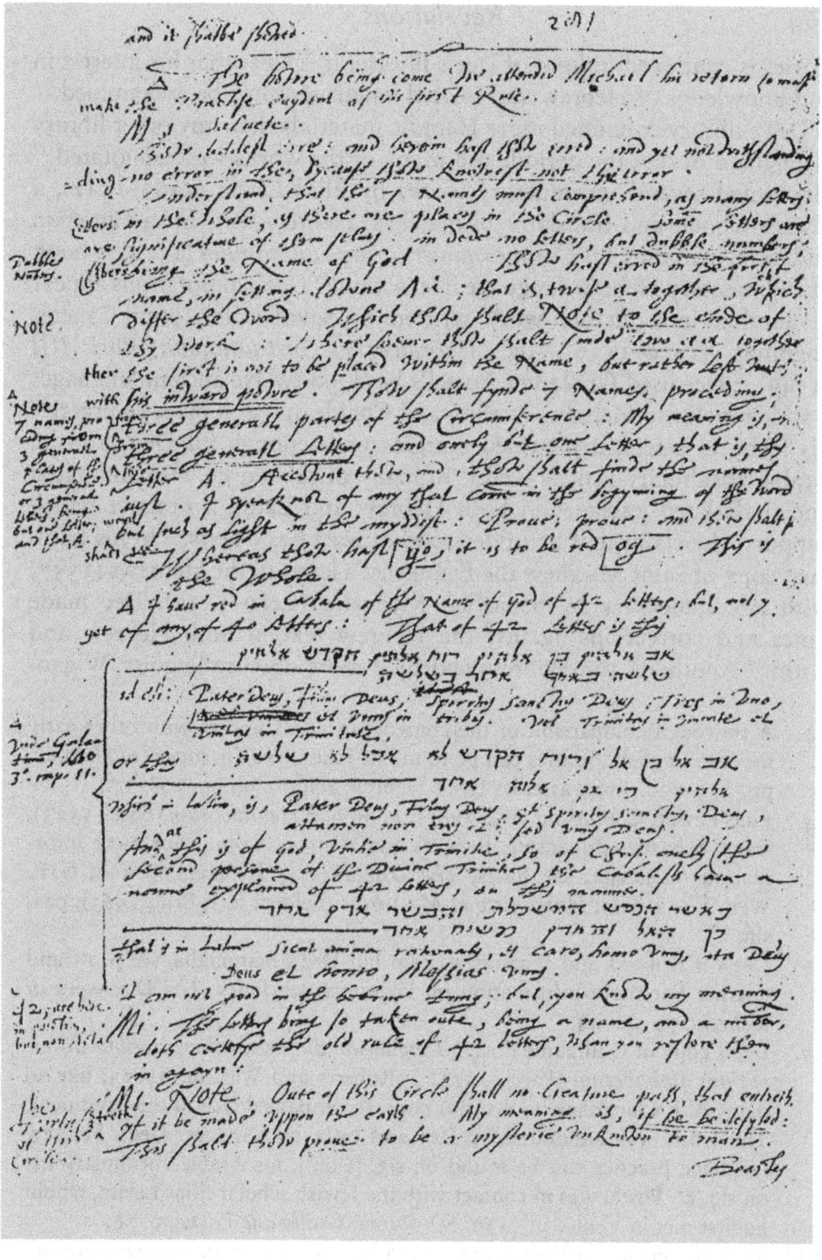

Figure 5. A page from the angel diaries showing Dee's command of cabala and Hebrew. Inscribed in the lower half of the page are the names of God in Hebrew. British Library Sloane MS 3188, f. 21ᵛ. Reproduced by permission of the British Library.

diaries is examined in light of Dee's library, it is clear that his interest in and knowledge of Hebrew and related languages must be reexamined.

Dee's library contained more Hebraic materials than any other library in England during the period, and many of the works were annotated.[24] He owned several introductory works by Elias Levita (1469–1542), a Jewish scholar and translator of medieval Hebrew texts for a Christian audience, including his *Composita verborum et nominum* (1525), a book of rudimentary Hebrew language instruction.[25] Dee's annotations appear in another introductory work written by Savonarola's pupil Sanctes Pagninus (1470–1536), the *Hebraicarum institutionum libri IIII* (1549).[26] Dee used Guillaume Postel's book on twelve ancient languages – which included Hebrew, Arabic, and Chaldean – as a place to practice his Hebrew letters; it provided margins wide enough for Dee to make Arabic vocabulary lists as well.[27] In many of the works, Dee annotated and underlined passages in both Hebrew and Latin, a practice which supports the claim that he understood at least basic texts.[28] In his bilingual copy of Saint Matthew the Evangelist's letter to the Hebrews (1557) with a commentary by Sebastien Muenster (1489–1552), Dee made notes and corrections against the Hebrew text in both Hebrew and Latin.[29] Another Latin-Hebrew text by a more obscure linguist, Wigan-

24.   A thorough comparison of the contents of Dee's library with other early modern English libraries can be found in Jones, *The Discovery of Hebrew*, pp. 275–290. Jones assesses Dee's Hebrew abilities on pp. 168–174.

25.   Elias Levita, *Composita verborum et nominum Hebraicorum* (Basel, 1525), trans. Sebastien Muenster, Roberts and Watson, #1594. For more information on Levita, see Jones, *The Discovery of Hebrew*, pp. 5–6; G. E. Weil, *Elie Lévita: Humaniste et Massorète* (Leiden: E. J. Brill, 1963), passim.

26.   Dee's is Roberts and Watson, #1569. For Dee's marginalia, see p. 15 and passim. For more information on Pagninus, see Jones, *The Discovery of Hebrew*, pp. 40–44.

27.   Dee's copy of Guillaume Postel, *Linguarum duodecim characteribus differentium alphabetum* (Paris, 1523) is Roberts and Watson, #1623; like so many of his language books, this copy accompanied Dee to the Continent. The work was annotated by Dee and by a later owner. Dee's Hebrew language practice can be found on sig. [Ciiii^v]; his Arabic vocabulary list on sig. E^v. Postel was in contact with the Jewish scholar Elias Levita, whom he first met in Venice in 1537. See Kuntz, *Guillaume Postel*, p. 26.

28.   See Dee's copy of Levita, *Composita verborum et nominum Hebraicorum*, sig. Cc4^r, for an example.

29.   Details about Dee's copy can be found in Roberts and Watson, #1588. For Dee's annotations, see the title page and passim. This volume was taken to central Europe in 1583. For more information on Muenster and his association with his teacher, Elias Levita, see Jones, *The Discovery of Hebrew*,

dus Hapellius, also contains marginal notes revealing Dee's interest in Hebrew.[30] From other manuscript marginalia we know that Dee felt sufficiently comfortable with his Hebrew to correct an exegesis of the tetragrammaton in his copy of Henricus de Herphs's *Theologiae mysticae* (1556).[31]

Although Dee's knowledge of Hebrew might have been poor by angelic standards or when compared to devoted Hebrew scholars like Reuchlin, it was perfectly adequate for understanding scriptural passages and the rudimentary principles of cabalistic exegesis. The angel diaries confirm that Dee was a highly knowledgeable Christian cabalist if not a highly proficient Hebrew scholar. Today an expertise in cabala is dependent upon a secure foundation of Hebrew, but the two were often mutually exclusive in the early modern period. Agrippa, for example, was not a proficient Hebraist, yet he was able to draw from the *Sepher Yetzirah* and Joseph Gikatilla's *Sha'are Orah* – neither of which was available in Latin. François Secret has pointed out that, in the case of Christian cabalists like Agrippa and Dee, limited language skills did not necessarily stand in the way of a knowledge of Jewish ideas and cabalistic techniques, even from obscure Hebrew texts.[32]

Except for the Hebrew language texts in his library, our knowledge of Dee's exposure to Jewish intellectual traditions remains speculative. We cannot rule out the possibility that he received formal training in Hebrew and cabala. Christian scholars interested in biblical antiquity frequently consulted Jewish instructors in the fifteenth century, as is indicated by Moshe Idel.[33] Rabbi Elijah Manahem Halfan commented on the continuance of this practice in the sixteenth century: "In the last twenty years, knowledge has increased, and people have been seeking everywhere for instruction in Hebrew." Christian students of Hebrew were seldom satisfied with language lessons alone but requested instruction in the cabala – a more problematic area of study, as Halfan explained, because among

pp. 44–48, and B. Barry Levy, *Planets, Potions and Parchments: Scientific Hebraica from the Dead Sea Scrolls to the Eighteenth Century* (Montreal: McGill-Queen's University Press, 1990), p. 25.

30. Dee's copy of Wigandus Hapellius, *Linguae sanctae canones grammatici* (Basel, 1561), contains annotations on pp. 59–60; Roberts and Watson, #1600. This work remained in England in 1583.

31. Dee's copy is Roberts and Watson, #223; his annotations appear on p. 9ʳ. Dee took Henricus's work with him to central Europe in 1583.

32. François Secret, *Les Kabbalistes chrétiens*, pp. 53–55.

33. Moshe Idel, "The Magical and Neoplatonic Interpretation of the Kabbala in the Renaissance," trans. Martelle Gavarin, in *Jewish Thought in the Sixteenth Century*, ed. Bernard Dov Cooperman (Cambridge, MA: Harvard University Press, 1983), p. 186.

the Jews there were "but a small number of men learned in this wisdom, for after the great . . . troubles and expulsions, but a few remain."[34] Dee might have sought out one of these scarce individuals in Louvain, where he purchased most of his Hebraica in the 1560s. Louvain was well known as a center of language study, and the university's trilingual college would have given Dee access to a number of proficient Hebraists.[35]

An advanced knowledge of Hebrew would have been of limited use in the angel conversations, however, because Dee was receiving God's own language rather than the decayed terrestrial tongue. The angels began delivering characters, words, and names in the divine language in March 1582 when Dee began his relationship with Edward Kelly.[36] The basis of the divine language was an alphabet of twenty-one regular characters and one aspirated character.[37] Here Dee would have seen a similarity between the divine language of God and contemporary Hebrew, since the Hebrew alphabet contained the same number of characters.[38] In addition, each character of the divine language also bore a name. When the characters were strung together to form words, the words were – like Hebrew – to be read from right to left.[39] Later, when words and phrases were drawn together into revealed books, they were arranged like a Hebrew text, with the first page of the book at the back of the volume and the final page at the front.[40]

The divine language's alphabet related only superficially to Hebrew, however, and even less to other known languages. The attempts of several scholars to ground Dee's angelic alphabet in various ancient, medieval, and early modern sources have met with mixed results. The most provocative and unsubstantiated suggestion concerning its origins was made by Donald Laycock, who argued that Dee may have been familiar with an Ethiopic version of the Book of Enoch, which most scholars believe reentered the Western tradition in the eighteenth century after disappearing for over a thousand years.[41] If this argument were proven, it would help to explain both the divine alphabet, which resembles Ethiopic in some respects, and Dee's angelology, for the extracanonical Book of Enoch had a rich and diverse assortment of angels. It is also

34. Ibid.
35. Jones, *The Discovery of Hebrew in Tudor England*, pp. 99–100 and 254.
36. Dee, *AWS* II:41–62.   37. Ibid., p. 237.
38. Whitby's discussion of the language and its relation to Hebrew and other languages appears in Whitby, *AWS* I:145–149.
39. Dee, *AWS* II:398.   40. Whitby, *AWS* I:144.
41. Laycock, *The Complete Enochian Dictionary*, p. 14.

possible that texts and orally transmitted knowledge about the prophet were circulating in early modern Europe; Guillaume Postel, for example, consulted an Ethiopic priest about the possible contents of Enoch's lost book earlier in the sixteenth century.[42]

However, we need not rely on Dee's possession of an unrecorded copy of the Ethiopic book of Enoch to explain his interest in the divine alphabet, for his extensive linguistic studies would have provided him with sufficient provocative information about God's "lost" language. A more likely source for the angels' divine alphabet is Giovanni Pantheus's *Voarchadumia contra alchimiam*, which Dee owned and annotated.[43] Pantheus's work contained an alphabet labeled "Enochian" that strongly resembles Dee's divine script. Dee's seventeenth-century editor, Meric Casaubon, believed that the characters of the divine language derived from Theseus Ambrosius's *Introductio in Chaldaicam linguam, Syriacum atque Armenicam et decem alias linguas* (1539). Although Casaubon might have based his argument on his own careful study of ancient languages, Dee did own a copy of Ambrosius's work.[44] Agrippa's *De occulta philosophia* also contained ancient-looking alphabets derived from Hebrew that would have supported the purported antiquity of the angelic alphabet revealed to Dee.

When Dee received the divine language from the angels he was not receiving a musty lexicon and grammar. Instead, he was given a living language that promised to reconnect parts of the cosmic system. Learning, speaking, preserving, and ultimately mastering the creative power embedded in the language provided Dee with a new course of study that he felt would take him toward wisdom. The language also provided him with an *activity* essential to the long-awaited restitution of the Book of Nature: a restoration of the lost communicative links between Adam, the angels, and God. This set Dee's study of the divine language apart from his earlier studies of ancient languages. Every mistake he made while learning the language took on cosmic significance.

Unfortunately, the divine language did not conform to any known rules of grammar, syntax, or pronunciation, and Dee was faced with a difficult task when it came to learning its basic principles. He paid scrupulous attention to the proper spelling and pronunciation of every word the angels revealed, though he was frustrated by the enormous

---

42. Kuntz, *Guillaume Postel*, p. 65.
43. Pantheus, *Voarchadumia* (Roberts and Watson, #D16), pp. 12–15.
44. Roberts and Watson, #1620. It has not been definitely recovered, although a copy in the Bodleian Library bears annotations in a hand that may be Dee's.

variations contained in the language. Dee's humanistic training was more a hindrance than a help in this process as each of the angel's speeches was minutely recorded and, later, reconciled to earlier diary entries. He worried about each discrepancy, repeatedly asking the angels for clarification. He was not reluctant, for example, to note, "I suspect this to be an imp[er]fect phrase" against dubious passages.[45] Many passages in the divine language were recorded along with marginal notes indicating English words that might help Dee with the erratic pronunciation schemes, as when he noted that the ending for "lusache" should be pronounced "as *che* in che[r]ry."[46] Spelling, or the "true orthography" as Dee called it, provided further difficulties, and he expressed his consternation over the "diverse sownds" accorded to the letters "g," "c," and "p."[47] For Dee, a scholar educated with humanistic standards of linguistic accuracy and suffering under the pressure of trying to learn a sacred language on which the future of the world depended, the lack of a systematic foundation for the divine language was intensely frustrating.

Dee spent most of his energy trying to master the way the divine language was to be spoken. The angels told him that this was crucial because it would render his speech unlike that of other mortals and more like the voices of angels.[48] The divine language would enable Dee "to talk in mortal sounds with such as are immortal."[49] As with so many aspects of Dee's angel conversations, there are provocative connections between these features and Dee's knowledge of Jewish beliefs. For Reuchlin and other cabalistic scholars, the real power of the tetragrammaton resided "not only [in] the characters and the word, but also [in] its pronunciation, which is occult and hidden."[50] According to the scholars, only Moses had been instructed in the proper way to vocalize the tetragrammaton, so Dee's preoccupation with the pronunciation of the divine language is understandable in light of his desire to follow in the footsteps of biblical sages and prophets.

Dee's efforts to learn the pronunciation rules, not to mention the grammar and syntax, for the divine language was further complicated by the time-consuming methods the angels used to convey God's language. Characters, phonetic pronunciations, words, and phrases all had to pass through the filter of the showstone and the scryer before they were avail-

---

45. Dee, *AWS* II:213.   46. Ibid., p. 300.   47. Ibid., p. 309.
48. Ibid., p. 268.   49. *TFR*, p. 66.
50. Charles Zika, "Reuchlin's *De Verbo Mirifico* and the Magic Debate of the Late 15th Century," *Journal of the Warburg and Courtauld Institutes* 39 (1976): 127.

able to Dee. One of the most common ways in which the angels conveyed a text was to appear standing in the showstone on a table covered with letters (Figure 6). The angels would point to each individual letter of the lesson with a long rod, spelling out entire prayers, one character at a time. It is not surprising that angel conversations concerned directly with the divine language could last over seven hours. As the hours and days stretched on, Dee's frustration mounted. Finally, he dared to ask whether the angels could not devise some better and more speedy way to deliver their language. After all, Dee argued, it was surely not in God's best interest for the process to take so long. The angels scolded Dee for his presumption and demanded that he learn the language on God's terms, not his own.[51] But Dee's temerity did lead the angels to modify their methods. As a concession to his human frailty, the angels began to convey long prayers by recitation, which Dee was permitted to record phonetically.[52] But the additional time Dee won through the speedier transmission was lost when he realized he then had to translate the prayers twice: first into the divine characters, and then into English.

Dee was not a model pupil in the angels' "celestial school," and the angels often took him to task for what they perceived as his lack of devotion to linguistic matters. Dee complained when the angels chided him, replying that he "had diverse affayres which at this present do withdraw me from peculier diligence using these Characters and theyr names."[53] Exceptions, muddled "orthography," and a complete absence of systematic pronunciation must have made Dee despair of ever understanding how the divine language functioned. The angels were initially sympathetic to his plight, but their patience soon expired. They told Dee that *"the ende and Consummation of all thy desired thirst"* for knowledge would result from his mastery of the language and not from the direct question-and-answer method that he preferred.

Dee believed that the divine language represented the final step in his climb to divine wisdom because of its initial infusion into the Book of Nature at the moment of Creation. Though some early modern scholars believed that God created the cosmos by use of the Word and shaped the first human, Adam, from the substance of the created world, the angels told Dee that the divine language played a more comprehensive role. As they recounted it, the divine language was the shaping or "plasmating" force for both the Book of Nature and humanity.[54] Thus it was

---

51. Dee, *AWS* II:267. 52. Ibid., p. 268. 53. Ibid., p. 242.
54. *TFR,* p. 371. I am indebted to Brian Copenhaver for explaining the Greek origins of the word *plasmation* and the use of the term *plasso* in the Bible.

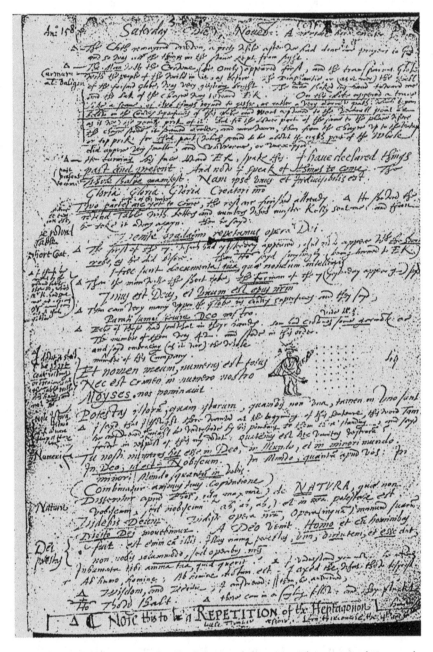

Figure 6. In this entry, Dee sketched the angel Carmara. The entry is also a good example of Dee's practices of marginal notation and underlining, which he used in both his library books and his angel diaries. British Library Sloane MS 3199, f. 49ʳ. Reproduced by permission of the British Library.

the divine language which lent coherence and structure to the entire creative act.

Given the prominent place of the divine language in ordering the Book of Nature, Dee believed that it contained a great deal of power. Because God and the Word were so intermingled in the early modern period as to be almost indistinct, the divine language was thought to contain not only the creative, ordering principle of the cosmos but also the power to avert evil and the power over life and death. Religious authorities and natural philosophers agreed that the search for God's own language was a fruitful and valuable intellectual project. Pope Leo X (1475–1521), for example, was reputed to have received an angel who brought a sample of holy writing from heaven. Pope Leo, in turn, gave it to King Charles V of Spain to serve as a protective talisman against evil.[55] The pope's interest in the divine language and his belief in its mystical power of language to protect the pious must have been well known at the time; Giovanni Pantheus's study of the relationship between language and alchemy, which so interested Dee, was dedicated to Leo X.[56]

Much of the divine language's innate power depended on how it was used. The primary reason Dee wanted to learn it, after all, was to communicate with both the terrestrial and the celestial levels of the cosmos with a minimal interpretive gap between God's Word and his own limited human understanding. Dee wanted to serve as a bridge between humanity and divinity, to act as a mediating angel conveying God's messages to the waiting world. But the divine language could not make Dee or his associates equal to the angels. Because the angels' powers and abilities were linked to their familiarity with the "Powre of the primitive divine . . . or Angelicall speche," Dee understood that his own skill with the language would always be a poor reflection of theirs.[57] Though the angels told Dee that his use of the divine language would make him "*differ . . . in speche frome a mortal Creature,*" they never suggested how this difference would be manifested, or to what extent Dee would be exempted from human limitations.[58]

Dee's angel diaries show no sense that the divine language was a "dead" language limited purely to textual methods of transmission. Instead, it was a living language that humanity had lost but might one day reclaim. Dee was sensitive to both the spoken and written power of the divine language and remarked that he was receiving both the "voice" and the "language" of God and the angels – the language in both its

---

55. "Magical Papers," British Library Add. MS 36674, ff. 65ʳ–91ᵛ, f. 87ʳ.
56. Thorndike, V:535–540.   57. Dee, *AWS* II:233.
58. Ibid., p. 268; Dee's emphasis.

spoken and written forms.[59] The proper pronunciation of the divine language was crucial because of its role in communication between the celestial and terrestrial parts of the natural world. The divine language had once been a crucial link between disparate levels of the cosmic system, and in the early modern period it had the potential to reinstate communication between God, the angels, humans, and the Book of Nature. Agrippa negotiated his way through pronunciation problems by arguing that the divine language was not, and never had been, vocalized like human speech. Rather, Agrippa believed that the divine language could pass from God to humanity, and then from mind to mind, through the imagination.[60] Because human imaginations were too weak to enjoy such communication in their present, decayed condition, only the angels and God still used the divine language to communicate. Guillaume Postel agreed that Adam had not needed to rely upon hearing God actually speak, but instead received the divine language directly into his mental faculties.[61]

Dee did not envision entirely mystical applications for the divine language that would involve a complete withdrawal from terrestrial affairs. Despite his special relationship with God and the angels, he remained committed to pursuing a knowledge of "the creation and created things" that could be shared with humanity.[62] Employing the revealed, divine language Dee continued to study the natural world. He believed that a combination of the divine language and natural philosophy would unlock the secrets of the created world. Cabala provided a way of uniting language and natural philosophy, and it was through a reformed cabala that Dee found a way to apply the divine language to the Book of Nature.

Restoring the Cabala

A full exploration of the intricacies of the cabala is beyond the scope of this study, yet some familiarity with this mystical system is necessary to an understanding of Dee's "cabala of nature." The cabala, or *Kabbalah* as it is known within the Jewish tradition, was a medieval mysticism that developed in the twelfth and thirteenth centuries and continued to

59. Ibid., p. 233.   60. Agrippa, *DOP* III:413.
61. For a discussion of Postel's ideas on this subject, see Kuntz, *Guillaume Postel,* pp. 38–39.
62. Dee, *AWS* II:118.

grow and change throughout the early modern period.[63] Like many other forms of mysticism, Jewish and Christian, the key cabalistic text was the Bible. In addition, cabalists relied on Rabbi Moses de Léon's *Zôhar* and other medieval commentaries for guidance and instruction. While other forms of Jewish mysticism, including *Hékhalot* and *Merkavah* mysticism, relied on direct revelation to supplement and enhance biblical teachings, the cabala was based on the interpretation of ancient and biblical texts through an exegesis of words, phrases, and symbols.[64] Cabalistic exegesis was never strictly limited to written texts, however, but extended to other symbolic expressions.

For Jewish cabalists, the most important symbolic expression of God was contained in ten successive emanations known as the *sefirot*. Often depicted in the form of a tree, the ten *sefirot* were a microcosmic representation of the relationship between God and the Creation. In the cabala, emanation provided a mechanism of causation that allowed God to be present in everything yet still removed from the created world.[65] The light metaphysics of some medieval Christian scholars, including Robert Grosseteste, depended on divine emanation, as did the elaborate

63. Because Dee and the Christian authors he drew on for sources use the word *cabala* rather than *Kabbalah*, and because it has become common to use the latter designation only in relation to the Jewish mystical tradition, I have adopted the Christian spelling. The most important introductions to the cabala, with differences in intepretation and emphasis, are Gershom Scholem, *Kabbalah* (Jerusalem: Keter Publishing House Jerusalem, 1977); and Moshe Idel, *Kabbalah: New Perspectives* (New Haven and London: Yale University Press, 1988). Additional information about how cabala intersects with other forms of Jewish mysticism can be found in Gershom Scholem, *Major Trends in Jewish Mysticism* (London: Thames and Hudson, 1955). A brief yet thorough general discussion of Jewish mysticism, including the cabala, is included in Geoffrey Wigoder, ed., *The Encylopedia of Judaism* (New York: Macmillan Publishing Company, 1989), pp. 512–515. François Secret's work on Christian cabala is the most reliable, though dated, treatment of this important body of early modern philosophy. See *Les Kabbalistes chrétiens de la Renaissance* (Paris: Dunod, 1964) and *Le Zôhar chez les kabbalistes chrétiens de la Renaissance*, Etudes Juives, vol. 10 (Paris: Mouton, 1964). Additional insights into Christian cabala, though less reliable, can be found in J. L. Blau, *The Christian Interpretation of the Cabala in the Renaissance* (New York: Columbia University Press, 1965).

64. Wigoder, *Encyclopedia of Judaism*, p. 512.

65. See Idel, *Kabbalah*, pp. 112–122, for a discussion of the role of emanation in the process of creation and the sefirot.

angelologies with which Dee was familiar. The concept of emanation was also closely linked to linguistic theories of the Creation, because the Word emanated from God and, ex nihilo, formed the world. These ideas about emanation were probably influenced by Neoplatonic thought.[66]

In addition, the *sefirot* were linked to the Bible by an intricate and subtle series of correspondences, so that nearly each word in the Bible corresponded to one of the *sefirot*, increasing the exegetical potential of biblical texts.[67] God, divine revelation, the created world, and its micro-cosmic representation in the *sefirot* thus became a single, complex system for the cabalist to understand and master in order to ascend toward divine wisdom and the Godhead. Christian cabalist Johann Reuchlin figured the *sefirot* were a ladder to the heavens: "[the *sefirot* are] the ten rungs of the ladder on which we climb to know all truth, be it of the senses, or of knowledge, or of faith."[68] The *sefirot* also offered another possible use as intriguing as it was potentially dangerous. Just as a cabalist could use a combination of cabalistic techniques, God's revelations in scripture, and His emanations in the *sefirot* to climb to God, so the same path could be used to draw divine, celestial powers down into the sublunar world.

This strand of cabala, called "practical cabala" or "theurgic cabala," is similar to other types of spiritual magic in the medieval and early modern periods.[69] Though the *sefirot* provided an avenue for the descent of the celestial powers, the secret names of God and the angels provided the means by which the powers were called down into the sublunar levels of the cosmos. Thus the theurgic, practical branch of cabala relied on the invocation and manipulation of secret names to draw divine emanations into the natural world. The Christian West became attentive to these ideas with the publication of Pico della Mirandola's cabalistic theses in 1486, and they were subsequently developed in two ways: first, to reclaim and appropriate the cabala to prove that Jewish beliefs contained prophecies of Christ and Christian teachings; and second, to explore the cabala's esoteric and magical possibilities.

The angels' cabala of nature shared several features with the Jewish and Christian cabalas. These similarities must have pleased Dee, because

66. Scholem, *Kabbalah*, pp. 98–116.

67. Scholem, *Major Trends in Jewish Mysticism*, p. 209.

68. Johannes Reuchlin, *On the Art of the Kabbalah, De arte cabalistica* trans. Martin Goodman and Sarah Goodman, intro. by G. Lloyd Jones (New York: Abaris Books, 1983), p. 51.

69. Idel, *Kabbalah*, pp. 173–199.

he considered his cabala of nature the original, uncorrupt cabala from which the Jewish and Christian cabala derived. Like the latter, Dee's cabala was revealed wisdom. The art of cabala had first been given to humanity, as Reuchlin explained in the *De arte cabalistica*, through direct revelation.[70] Many centuries of human frailty stood between that time and the cabala in use in the sixteenth century. Dee, on the other hand, received a direct cabalistic revelation from God. Any similarities between the extant Jewish and Christian cabalas and Dee's cabala of nature, in his view, demonstrated the places where ancient systems had managed to retain their "true" features despite the vicissitudes of time and human understanding.

Cabalistic systems – Jewish, Christian, and even Dee's cabala of nature – depended on exegetical interpretations and the belief that language was instrumental in the Creation. But the language of Dee's cabala of nature was not Hebrew; it was the true, divine language God used in His first creative acts. Because the divine language was alphanumeric, like Hebrew, Dee believed that he could apply traditional cabalistic exegetical techniques to names, words, and phrases the angels conveyed. The angel diaries suggest that Dee used the three main cabalistic exegetical methods of *gematria*, *notarikon*, and *temurah*. *Gematria* was the most mathematical method, as it involved taking the numerical value of words and phrases and searching through scriptures for other words or phrases with a similar numeric value. In *notarikon*, the letters of a word were seen as abbreviations for entire sentences. The third cabalistic exegetical technique, *temurah*, transposed letters within a word to create new words.[71] These three cabalistic methods, as Reuchlin explained, ensured that no "word, no letter, however trifling, not even the punctuation, was without significance."[72]

References to the three methods appear throughout the angel diaries. Each letter of the angelic names inscribed on the Sigillum Dei signified another angel in a form of angelic *notarikon*.[73] In another example of *notarikon*, an angel explained to Dee that his words contained sentences of wisdom.[74] Dee also had "diverse talks and dyscourses of Transpositions of letters," or *temurah*, while engaged in the angel conversations, a method he had discussed earlier in his *Monas hieroglyphica*.[75] Like most cabalists, Dee believed the transposition of letters should follow

70. Reuchlin, *De arte cabalistica*, p. 51.
71. Scholem, *Major Trends in Jewish Mysticism*, p. 100.
72. Reuchlin, *De arte cabalistica*, p. 69.   73. Dee, *AWS* II:76.
74. Ibid., p. 224.   75. Dee, *MH*, p. 209.

certain rules, and he shared his "rule for to know certaynly how many wayes, any number of letters . . . might be transposed" with Edward Kelly during their conversations.[76]

The most complex incorporation of cabalistic exegetical techniques in the angel conversations occurred in Dee's use of *gematria*, the exegetical method that relied on numerical interpretations. *Gematria* depended upon the alphanumeric quality of the divine language. The angels Gabriel and Nalvage explained that "all the World is made of numbers," but the "Numbers we speak of, are of reason and form, and not of merchants."[77] The numerical quality of the divine language was not the mundane mathematics of Dee and his contemporaries, but was part of the world of divine essences and true forms. The angels' distinction resembled one made by Agrippa in his *De occulta philosophia*, where he differentiated between the "vocal numbers" used by merchants, and rational, formal, and natural numbers that contained the mysteries of God and nature.[78] Dee also adhered to this distinction between earthly and heavenly numbers. For him there were two types of number: "Number Numbyring," the numbers in the Creator, in spiritual and angelic minds, and in the soul of man; and "Number Numbred," which were found in all other parts of the created world.[79] Because numbers played a major role in both the cabala and in Dee's conception of nature, it is not surprising that he was interested in the numerological overtones of the divine language and the angels' cabala of nature. Numerology, though not strictly speaking cabalistic, could be related to Jewish mystical beliefs by Christian authors and might have gained an aura of extra authority because of this association. Descending from a rich assortment of ancient and medieval theories, the numerological beliefs of the early modern period were incorporated into architecture, music, and natural philosophy.[80] Numerological speculations derived from scripture were a feature of the most authoritative biblical commentaries and provided fuel for popular prophecies as well.[81]

Other references to cabalistic ideas and authors appeared throughout the angel diaries. Some references were no more than a notation of the

---

76. Dee, *AWS* II:351. 77. *TFR*, pp. 92–93.

78. Agrippa, *DOP*, pp. 368–369. 79. Dee, MP., sig. i^v.

80. See, for example, G. L. Hersey, *Pythagorean Palaces: Magic and Architecture in the Italian Renaissance* (Ithaca, NY: Cornell University Press, 1976).

81. See Vincent F. Hopper, *Medieval Number Symbolism* (New York: Cooper Square Publishers, 1969), for a discussion of these ideas.

word *cabala,* or Dee's notes to look for further information in a book by Reuchlin, Agrippa, d'Abano, or some other cabalistic authority. When the angel Galuah stepped forward in 1583 and told Dee that her garments were called "HOXMARCH," for example, he immediately associated the word with one of the ten *sefirot*: *Hokmah,* or divine, redemptive wisdom. The first of the divine emanations, *Hokmah* was the *sefirah* associated with the moment of God's initial conception of the Creation – the moment when all things remained perfectly conceived and contained by God.[82] Though Galuah's explication of the word's meaning has been obliterated in the manuscript, Dee's next remarks expressed his belief that the beginning of wisdom was found in the fear of God.[83] Even the description of the cabala of nature as a "three-fold art" capable of "ioyning Man (with the Knowledge *of the* WORLDE, the GOVERNMENT of his Creatures, and the SIGHT of his Maiestie" is reminiscent of the cabalistic "three-fold art" of *gematria, notarikon,* and *temurah.*[84]

Adam's Legacy: True Names and Lost Meanings

The divine language's ability to refer to objects in the natural world fully and completely was an important aspect of early modern interest.[85] This interest increased in the seventeenth century when debates began about the merits of creating a "universal language" that would reflect the essential consonance between human speech and the Book of Nature and reduce the confusion of tongues. Though artificial, this language was described as a "natural language" to distinguish it from the divine language used by God in the Creation and spoken by Adam.[86] Adam enjoyed a special level of communication with the Book of Nature which stemmed from one of his first acts in Paradise: giving each thing within

82. Scholem, *Major Trends in Jewish Mysticism,* p. 219.
83. *TFR,* p. 19. Dee's remarks were in Latin: "Initium sapientiae est Timor Domini."
84. Dee, *AWS* II:207.
85. See A. C. Howell, "Res et verba: Words and Things," *English Literary History* 13 (1946): 131–142, for a discussion of this idea in the early modern period.
86. See Vivian Salmon, "Language-Planning in Seventeenth Century England: Its Contexts and Aims," in *In memory of J. R. Firth,* ed. C. E. Bazell et al. (London: Longmans, 1966), pp. 370–397; and Thomas C. Singer, "Hieroglyphs, Real Characters, and the Idea of a Natural Language in English 17th Century Thought," *Journal of the History of Ideas* 50 (1989): 49–70.

the Book of Nature a true name that reflected its essential nature. As the angels explained to Dee, Adam "named all things (which knew it): and they are so in dede, and shallbe [sic] forever."[87] Dee, on the other hand, named the parts of nature incorrectly because his voice captured only a corrupt remnant of the divine language.[88]

The notion that words could refer to the essences of things has a long history not limited to discussions of Adam and the Creation. Plato's *Cratylus* suggested that language was capable of making words and their referents correspond exactly, though it was ambiguous regarding whether language was entirely natural or at least partially conventional.[89] D. P. Walker believed that the "verbal force" associated with the divine language of creation and other very ancient languages depended upon a linguistic theory that included "a real, not conventional, connection between words and what they denote."[90] Dee adhered to such a belief in his angel conversations and thought that Adam's language was capable of capturing the essences embedded in the Book of Nature.

Because Adam's true names had the potential to disclose the secrets of the Book of Nature, Dee was warned that he and his scryers must not recite the names and words given to them in the divine language as they were received. By working in this fashion of "dubble repetition," Raphael warned, "you shall both write and work all at ones [sic]: which mans nature can not performe." Instead, Dee was to wait until a time appointed by God to utter the words and phrases.[91] To circumvent the problems associated with Dee's having to hear, repeat, and write lengthy passages in the divine language, his angelic teachers began to utter their revelations in reverse order. When it was time for Dee to "activate" nature, God would instruct him to reorder the passages and unlock their dormant power. Dee was pleased by this method and understood its benefits, "else, all things called would appear: and . . . hinder our proceeding in learning."[92]

A knowledge of true names was also important to Christian and Jewish cabalists of the period, especially the true names of the angels, for they enabled a natural philosopher or cabalist to transcend the limits

87. Dee, *AWS* II:234.   88. Ibid., pp. 232–233.
89. See Brian Vickers, "Analogy versus Identity: The Rejection of Occult Symbolism, 1580–1680," in *Occult and Scientific Mentalities in the Renaissance*, ed. Brian Vickers (Cambridge: Cambridge University Press, 1984), pp. 97–100.
90. D. P. Walker, *Spiritual and Demonic Magic from Ficino to Campanella* (Notre Dame, IN: University of Notre Dame Press 1975), pp. 80–81.
91. Dee, *AWS* II:284.   92. *TFR*, p. 78.

of the decaying sublunar world and communicate with divinity. The true names for the angels appeared early in Dee's surviving angel conversations – even earlier than the characters of the divine alphabet, which we would consider a crucial starting point. For Dee and his contemporaries, however, beginning with the true names of God's angels was a more logical starting place, for it was through their mediating efforts that additional exposure to God's mysteries took place.

Several volumes in Dee's library were concerned with the secret names belonging to God, Christ, and the angels. Pico della Mirandola, in his cabalistic "Conclusions," argued that these names, especially the name of Jesus, contained powerful and persuasive messages about Christ's divinity and relationship to God.[93] In *De verbo mirifico,* Johannes Reuchlin built on Pico's theories in a treatise devoted to the marvelous effects that could be achieved with the names of Christ and other "wonder-working words" embedded in the Book of Nature. Reuchlin believed that these words were capable of restoring natural philosophy and would give human beings the same mastery of nature which Adam had enjoyed in Eden.[94] One such word was the tetragrammaton; two others, "Pele" and "NA" appear in the angel conversations, along with Dee's marginal notes regarding their explication in Reuchlin's *De verbo mirifico* and *De arte cabalistica.*[95]

Dee's interest in the power of names was part of his fascination with the occult properties of nature. Pico della Mirandola was one of the first to suggest the strong connections between magic, language, and nature, and his concept of a linguistic magic is often called "practical cabala" because of its active potential. Pico's ideas spread quickly through Eu-

93. Giovanni Pico della Mirandola, "Conclusiones," from *Opera omnia* (Basel, 1572), pp. 108–109. Dee owned a 1532 edition of the "Conclusiones," Roberts and Watson, #974. William Sherman believes that British Library 1606/318 was Dee's copy, based on the annotations. William H. Sherman, "A Living Library: The Readings and Writings of John Dee" (diss., University of Cambridge, 1992), p. 300.

94. Johannes Reuchlin, *De verbo mirifico* (Lugduni, 1552). Dee took his copy of the work, a Cologne edition of 1532, with him on his trip to central Europe and it has not been recovered. See also Zika, "Reuchlin's *De verbo mirifico* and the magic debate of the late 15th century," pp. 104–138, for a discussion of Reuchlin's ideas. For insights into the ways in which other European philosophers incorporated ideas about secret and true names into their work, see Brian P. Copenhaver, "Lefèvre D'Etaples, Symphorien Champier, and the Secret Names of God," *Journal of the Warburg and Courtauld Institutes* 40(1977): 189–211.

95. Reuchlin, *On the Art of the Kabbalah,* p. 111. Dee, AWS II:28–30.

ropean intellectual circles following the publication of his cabalistic *Con-clusiones*.[96] Reuchlin's *De verbo mirifico* placed a similar emphasis on the supernatural actions that certain words could produce.[97] Language was even more operative in Agrippa's *De occulta philosophia* because the magus's knowledge of secret, true names was required before any magical procedure could take place.[98]

From the angels. Dee received the true names of God, the names for multitudes of angels, the true names given by God to geographical regions of the earth's surface, the names of parts of the heavens, and the proper names for stages in the alchemical process. The secret names of God were first given to Dee when he received the Sigillum Dei design.[99] Later, on 21 March 1582, the angels gave Dee a tablet containing more arcane information: the *"Names of God, not known to the Angels"* which could not "be spoken or re[a]d of man."[100] Though Dee was able to record these names, the angels told him he would not be able to utter them until the final days, when they would elicit the appearance of the seven angels of the seven heavens closest to the earth, which *"stand allwayes* before the face of God."[101] Most angel conversations empha-

96. See Antonella Ansani, "Giovanni Pico della Mirandola's Language of Magic," in *L'Hebreu au temps de la Renaissance*, ed. Ilana Zinguer, Brill's Series in Jewish Studies 4 (New York: Brill, 1992), pp. 89–104; Dora Baker, *Giovanni Pico della Mirandola, 1463–1494, sein Leben und sein Werk* (Dornach: Verlag am Goetheanum, 1983); Guido Massetani, *La filosofia cabbalistica di Giovanni Pico della Mirandola* (Empoli: Tipografia di Edisso Traversari, 1897); Antonino Raspanti, *Filosofia, teologia, religione: L'unita della visione in Giovanni Pico della Mirandola* (Palermo: Edi OFTES, 1991); Chaim Wirszubski, "Francesco Giorgi's Commentary on Giovanni Pico's Kabbalistic Theses," *Journal of the Warburg and Courtauld Institutes* 37 (1974): 145–156; Chaim Wirszubski, *Pico della Mirandola's Encounter with Jewish Mysticism* (Cambridge, MA: Harvard University Press, 1988). For a modern edition of the *Conclusiones*, see Giovanni Pico della Mirandola, *Conclusiones sive theses DCCCC Romae anno 1486 publice disputandae, sed non admissae*, introduced and annotated by Bohdan Kieszkowski (Geneva: Librairie Droz, 1973).

97. See Zika, "Reuchlin's *De Verbo mirifico*," pp. 106–107, for a discussion of Reuchlin's ideas regarding the supernatural power of words.

98. Agrippa, *DOP* I:70.   99. Dee, *AWS* II:71–78.

100. Ibid., p. 93.

101. Ibid., 94. It would seem from this description that the seven angels mentioned were the angelic rulers of the "seven heavens," which would have been mentioned in the Sepher Raziel (an early medieval cabalistic work of great popularity), Peter d'Abano's *Heptameron*, and Agrippa's *De occulta philosophia*.

sized the names of angels, however, especially those who supervised the natural and supernatural worlds. Dee received, for example, the names of forty-nine separate angelic Governors who oversaw everything from princes and trade to the quality of the earth and the waters and the motions of the air.[102] Dee spent several days receiving these angels' names and grouping them into seven orders of seven.

Dee explored the exegetical potential of the true names of God and His ministering angels whenever they were revealed. When the angels gave Dee the angelic names to inscribe on the Sigillum Dei, he was told that each letter of each angel's name could bring forth a "daughter," and each daughter's name could bring forth another daughter.[103] This method of nominal exegesis was derived from *notarikon* and had been used by other cabalistic scholars. Reuchlin explained in *De arte cabalistica*, for instance, how the four letters of the tetragrammaton could yield seventy-two angels "as if by divine issue," demonstrating once again that ideas in Dee's angel conversations are not entirely idiosyncratic.[104] Through *notarikon*, Dee received an intricate linguistic genealogy of angels, though it is not evident whether he subjected other angelic names to similar cabalistic processes.

The Cabala of Nature

The exegetical potential and magical power of true names represented the first step in a much larger exegetical task: deciphering the Book of Nature. A proper exegesis of the Book of Nature was crucial to Dee's natural philosophy. The Jewish cabala was intended for scriptural exegesis, not for an exegesis of nature, a distinction that problematized Dee's use of cabalistic techniques to explain and understand the world. These difficulties did not, however, prohibit natural philosophers like Dee or Giovanni Pantheus from exploring the possibilities of a cabala suitable for the Book of Nature. Dee's *Monas hieroglyphica* and Pantheus's *Voarchadumia contra alchimiam* demonstrate how cabalistic exegetical techniques were applied to interpretive problems in the natural world.[105] Even so, the Jewish cabala and the natural world did not interact as neatly as many natural philosophers wished. As with so many

102. Dee, *AWS* II:122.   103. Ibid., pp. 76–91.
104. Reuchlin, *De arte cabalistica*, p. 267.
105. An analysis of the cabalistic overtones in Dee's *Monas hieroglyphica* can be found in Michael T. Walton, "John Dee's *Monas hieroglyphica*: Geometrical Cabala," *Ambix* 33 (1976): 116–123.

of Dee's intellectual difficulties, the angels provided him with a way out of this dilemma. They did so by revealing a body of information which they referred to as the "true cabala of nature." Dee was intrigued by the possibilities offered by this interpretive art, which depended on his grasp of the divine language and Adam's true names. Understanding the place the cabala of nature occupied in Dee's natural philosophy and theology brings a modern reader to the heart of his angel conversations and to the reasons why he was so consumed by their revelations.

First, it is important to clarify that the angels' cabala of nature was not a direct derivative of the traditional Jewish cabala, or even the Christian cabala. Though the cabala of nature is provocatively reminiscent of many basic tenets of cabalistic study, because the cabala of nature was intended to disclose the secrets of a different "text" – the Book of Nature rather than the Book of Scripture – its exegetical techniques were significantly different. If we see the cabala of nature as analogous to, but not dependent upon, the Jewish and Christian cabala, it becomes possible to examine Dee's cabala of nature on its own terms. As a result, we gain an understanding of the cabala of nature both as it distantly related to the Jewish and Christian cabala, and as a system of exegesis in its own right.

Given the scholarly dismissal of the angelic language as a hodgepodge of unutterable sounds and a strange assortment of names and phrases, it is understandable how the angels' "true cabala of nature" might also be consigned to virtual oblivion. Fragmentary because of the loss of sections of the angel diaries, seemingly devoid of a systematic foundation, and delivered in a scattered, piecemeal fashion, the cabala of nature is frustrating to study. That Dee himself was usually confused about the precise applications of the art and its intricate methods is little consolation. Even its designation as a "cabala" of nature has done little but complicate the situation, since an exploration of the works of other early modern cabalists has failed to uncover useful comparative information.

Dee's cabala of nature was based on the interplay between the elemental and spiritual numbers of the numerological tradition: four (signifying fire, air, earth, and water) and three (signifying the holy trinity of Father, Son, and Holy Spirit), respectively. Alone and in combination, these numbers governed the Book of Nature's cabalistic system. The most important combination of the two numbers in the cabala of nature was seven, which signified a combination of the elemental and the divine. As the angels explained to Dee:

> Seven comprehendeth the Secrets of Heven and e[a]rth: *seven knit-teth mans sowle and body togither (3, in sowle,* and *4 in body)* In 7, thow shalt finde the Unitie: In 7, thow shalt finde the Trinitie[.]

In 7, thow shalt finde the Sonne, and the proportion of the Holy Ghoste.... Thy Name (O God) be praysed for ever, from thy 7 *Thrones,* from *thy 7 Trumpets,* and from *thy 7 Angels.... In 7 God wrowght all things. In 7, and by 7 must you work all things.*[106]

The emphasis on the number seven was not unique to Dee's angel conversations. Seven occupied a prominent place in biblical commentaries because of the frequency with which it appeared in Revelations, the final book of the Bible. In Revelations 8:2, for example, seven thrones, trumpets, and angels warned humanity of the coming Apocalypse. Dee's cabala of nature included many permutations of seven, such as forty-nine angelic Governors and their "calls," and forty-nine gates of nature and their "keys." The cabala of nature also included thirty Airs ruled by twelve angels, and forty-nine-page books – and these numbers can be obtained by adding and multiplying the numbers three and four in various combinations.[107]

Dee's cabala of nature can be divided into two branches: one primarily concerned with drawing celestial influences down into the natural world; the other primarily concerned with restoring the communication between humanity, divinity, and the Book of Nature. Each facet of the cabala of nature was angelically delivered, followed by a short pause while Dee digested the new information, asked questions, and received angelic tutoring. Then the delivery of the cabala of nature resumed. While both branches of the cabala of nature were parts of a single system, they could also stand separately.

The first branch, like theurgic or practical cabala, promised to enable Dee to draw celestial influences into the sublunar natural world. Referred to here as the "angelic" system of the cabala of nature, this branch consisted of the secret, true names of the "Sons and Daughters of Light" and the "forty-nine angels who govern the earth." Both the Sons and Daughters of Light and the angelic Governors had specific roles in nature but were not present in the terrestrial world. Dee needed to employ special "calls" to bring the angelic Governors into the natural world to do God's will.[108] This angelic system was articulated at an early point in the Dee–Kelly relationship, and remained a part of the angel conversa-

---

106. Dee, *AWS* II: 80–81; Dee's emphasis.
107. The thirty Airs [$(3+4+3)\times 3$], ruled by twelve angels [3×4], and forty-nine page books [$(3+4)\times(3+4)$].
108. The angelic calls are *not* the "claves angelicae" revealed to Dee in Cracow in 1584, now British Library Sloane MS 3191. Rather, the "claves angelicae" were the "keys" to open the Gates of Nature and were part of the second, apocalyptic branch of the cabala of nature.

tions, because the large wax seal, or Sigillum Dei, was inscribed with the names of the Sons and Daughters of Light.

The role of the Sons and Daughters of Light in the cabala of nature is ambiguous. Typically, Dee was introduced to each angel, given his or her proper name, and instructed concerning their duties and responsibilities. In the case of the Sons and Daughters of Light, though, the possible uses for the angels were seemingly indefinite, as was their number. Successive generations were described in a complex and confused fashion, gathered into imprecise genealogies of "Sons of the Sons of Light" and "Daughters of the Daughters of Light," and so forth. A total of twenty-eight Sons and Daughters of Light was specifically mentioned, but the later generations were not introduced fully to Dee; not even their names were given to him. The names of the Sons and Daughters of Light were inscribed on the Sigillum Dei, and one of their roles was to facilitate the manifestation of the angels in the showstone. This made sound metaphysical sense, since all angels were creatures of light, and the Sons and Daughters of Light were tied nominally to God's emanations. Other functions for the Sons and Daughters of Light were less clear. Seven of the Sons of Light appeared carrying metal balls of gold, silver, copper, tin, iron, mercury, and lead – but the angels did not tell Dee whether the balls were meant to signify only the metals, or other features of the natural world associated with them, such as the planets.[109]

More integral to Dee's cabala of nature were the forty-nine angels who governed the earth. These angels were the cornerstone of Dee's angelic cabala, and were linked to the divine language, the natural world, and the cabala of nature in many ways and on many levels. First, through the Governors, Dee gained a valuable, "angelic" perspective on the natural world, a point of view that was above and beyond immediate sensory experience of the terrestrial world.[110] Second, the names of the Governors make an explicit reference to the natural world and the divine language that created it.[111] Each Governor's seven-letter name began with a "B" – the first letter of the divine language, the first letter of the Hebrew Genesis, and the letter that began God's creative process in the cabalistic text *Zôhar*.[112] The cabalistic belief in the power of names and the potential for names and words to disclose secret and profound infor-

---

109. The latter is more likely, as the planets were celestial bodies that were thought to be made of light – there was no sense in the sixteenth century, as there is today, of the planets as solid bodies. Dee, *AWS* II:88–89.
110. Ibid., p. 119.
111. The conversation took over seven hours. It appears in ibid., pp. 102–130.
112. Whitby, *AWS* I:125.

mation was present in Dee's understanding of the Governors' names. Uriel told Dee that, within their names, "The Fontayne of wisdome is opened. Nature shalbe knowne: Earth with her secrets disclosed. The Elements with theyr powres judged."[113]

The Governors were grouped into seven orders of seven angels. This arrangement numerologically underscored their significance in the elemental world created in seven days, and in the heavenly world of the seven planets. Each of the seven orders governed a specific feature of the created world. The first order governed wisdom; the second, earthly kings; the third, the waters; the fourth, the earth; the fifth, living creatures; the sixth, the motions of the air; and the seventh order was responsible for banishing evil angels and spirits. These tasks were more complex than simply overseeing some part of nature, for the forty-nine angels governed the intangible (wisdom) as well as the tangible (the earth and living creatures). The seven orders of Governors were not democratic, with equal powers enjoyed by every angel. Rather, they were arranged hierarchically, with one angel designated as the king and another as the prince; the responsibilities of the five remaining angels in each order were not clearly stated, or have not survived.

Within the orders, the duties of the king and the prince differed. The kings had jurisdiction over a specific aspect of the world while maintaining an overall responsibility for everything the order governed. The princes had more specific and limited duties. While King Bnaspol governed the earth, for example, his prince, Blisdon, was particularly responsible for the subterranean. The king governing wisdom, Bobogel, had under his subjection Prince Bornogo, who knew how to "make nature perfect" and about the earth's minerals and metals – in other words alchemy.[114] Though the relationship between kings and princes was logical, the delivery of this information was not. The angels first revealed the kings and their responsibilities, then the princes. This caused some confusion when Dee tried to draw the system together into a coherent whole. Finally, however, Dee was able to place the names of the angels into a circular wheel, similar to Lullian and cabalistic wheels, and clarify their relationships.[115]

An angel's visual appearance could provide valuable clues to his role and responsibilities, and the Governors were no exception to this rule. Their appearance was especially helpful to Dee, given the confusion surrounding their duties. The seven kings carried emblems of their pow-

---

113. Dee, *AWS* II:124.   114. Ibid., pp. 141 and 147.
115. Ibid., p. 129. For the wheels' possible relationship to cabalistic wheels in Rabbi Akiba ben Joseph's *Book of Formation*, see Whitby, *AWS* I:124.

ers, which Dee faithfully noted. These emblems exemplified their function in the cabala of nature; thus the angel in charge of wisdom held books while the angel of the earth carried herbs.[116] Princes who served under the Governor of wisdom, King Bobogel, appeared as "Auncient and grim Cowntenanced men in black gownes" holding books, "and some had stiks like Measures," which Dee thought might relate to an expertise in geometry.[117] Though the angels referred to them as the "Princes of Nature," both their clothing and their place in the angelic order indicated that they were connected to wisdom, and Dee noted in the margins "philosophers," to describe their appearance and the role that their appearance reflected.

Dee was preoccupied with his ability to communicate with the angelic Governors and how he might benefit from their knowledge of the natural world. To resolve these concerns, the angels gave Dee forty-nine "calls" (sometimes referred to as "claves" or "keys") between 31 March 1583 and 6 April 1583.[118] The angelic calls were contained in a separate book, except for one call not known by Dee or anyone else save God.[119] The angels conveyed the calls in the divine language, character-by-character, which took a great deal of time, especially as Dee and Kelly were too distracted with other matters to master the intricacies of the alphabet.[120] At last the angels made a concession to Dee's other responsibilities and began to recite the calls in the divine language, but even so the delivery, recopying, and discussion of the calls took nearly a month. Today, forty calls survive in the divine language. The angel diaries suggest that the final eight calls were received and placed "in another boke," possibly with an English translation, but they have not been located.[121]

The angels promised Dee that he would one day be able to use the calls to resolve the corruptions of the Book of Nature. With the help of the angelic Governors, he would "work in the quieting of the estates, In *lerning* of wisdome, pacifying the Nobilities judgement . . . as well in the depths of waters, Secrets of the Ayre, as in the bowells and entrails of the Earth." Dee could not use the calls until God deemed it was time, however. Until that time, Dee could only gain the support of the Gover-

---

116. Dee, *AWS* II:123–124.   117. Ibid., p. 140.
118. Ibid., p. 263–321.   119. Ibid., p. 227.   120. Ibid., p. 259.
121. Ibid., p. 320. Dee might have destroyed it when thirty angelic books were thrown into a furnace in Prague at the angels' command, though he later claimed that all materials had been restored except for the angel conversations with Francesco Pucci. It is likely that the English translation was lost in the bottom of the London confectioner's pie plates. *TFR*, p. 418 and Dee in Josten, p. 249.

nors by *"Invocating uppon the Name of GOD"* through prayer.[122] Even
then, the Governors would act only according to God's will, which could
be withdrawn if Dee did anything contrary to His wishes.[123] In the cabala
of nature, as in every other aspect of the angel conversations, Dee was a
supplicant requesting that God use him for His purposes, rather than a
magus manipulating the forces of nature.

The angelic system of the cabala of nature with its angels of Light and
Governors of the Earth, despite promises of information and fascinating
names, was soon put aside in favor of the earthly patronage opportunities afforded by the arrival of Count Laski. Not until Dee and his
household joined Laski and moved to Cracow in 1583 did the cabala of
nature return to the forefront of the angel conversations, this time with
an increasingly apocalyptic tenor and purpose. The conversations became more apocalyptic when the angels instructed Dee to move from
Cracow to Prague in July 1584 and used imagery which reinforced Dee's
belief that the final days were imminent. The cabala of nature responded
to the change in mood. Endless catalogues of angelic names and angelic
roles in the Book of Nature were replaced by fearsome prayers that
called upon entirely new and unfamiliar angels who were to play a part
in the Day of Judgment.

The apocalyptic system of the cabala of nature was based on the forty-
nine Gates of Nature and their "claves" or keys, the thirty Airs governed
by the angels of the twelve tribes of Israel, and the Book of Enoch, which
contained "the Invocation of the names of God and of the angels by the
names of God."[124] The angels emphasized Dee's *use* of the information
contained in this second branch of the cabala of nature. Still, there were
many similarities between the angelic and apocalyptic branches of the
cabala of nature, both in the methods of delivery and in the way that
the systems were structured.

The angels conveyed the Gates of Nature to Dee between 13 April
1584, and 14 May 1584 while he was in Cracow. From the onset the
angels were unusually forthcoming about Dee's use of the Gates: their
knowledge would enable Dee to "judge, not as the world doth, but
perfectly of the world, and of all things contained within the Compasse
of Nature, and of all things which are subject to an end."[125] These results

---

122. Dee, *AWS* II:125; Dee's emphasis.
123. Though Whitby believes that the "forty-nine governing spirits make the
      magical system [of the angel conversations] daemonic in a very practical
      way" this is not the case, for the angels' actions did not hinge upon Dee's
      abilities or desires. Whitby, *AWS* I:129.
124. *TFR*, p. 189.   125. Ibid., p. 77.

are significantly different from those described in the angelic branch of
the cabala of nature, for while Dee's knowledge of the angelic Governors
would help him shape the policies of the nobility and further human
understanding of the Book of Nature, his knowledge of the Gates would
increase his powers of discernment and his ability to *judge* aspects of the
natural world. Within the Gates, the angels told Dee, he would discover
the "Dignity and Corruption of nature," and would be made aware "of
the secret Judgments of the Almighty."[126] Nalvage told Dee that the keys
to the Gates of Nature contained the "mysteries and secret beings and
effects of all things moving, and moved within the world."[127] Dee was
to receive the information in two ways: secret judgments were to come
immediately from God, as they did to the Apostles; and information
about nature was to be mediated through angelic revelations. The angel
Nalvage, Dee's tutor in the intricacies of the apocalyptic system, told
Dee that he would share forty-eight keys that would open nature and
the gates of understanding.[128] As in the case of the forty-nine Governors
(one of whom could not be summoned), one of the Gates of Nature was
to remain closed for God's use.

The "keys" were revealed by Nalvage, who appeared in the showstone
bearing a rod and poised within a grid of alphabetic tables. He then
pointed from letter to letter with his rod, spelling out each word and
phrase of the keys.[129] Nalvage shared the exact location of every character in the tables, and Dee noted this information in puzzle-like formulas of ascending and descending squares. Because the angel diaries do
not contain an explanation of how these tables related to the Gates, it is
difficult to grasp why Dee found the locations of the letters so intriguing.
One possible interpretation lies in the descriptions of the tables revealed
a few days before the keys. Nalvage showed Dee several different tables
beginning on 10 April 1584, including a "round Table of Christal, or
rather Mother of Pearl" that had "an infinite number of letters . . . as
thick as one can stand by another" on its surface.[130] The round table
represented "the beginning and ending of all things," and Nalvage attributed the substance of the table to God the Father, its circumference
to Christ ("the finger of the Father, and mover of all things"), and "the
order and knitting together of the parts" of the table in their "due and
perfect proportion" to the Holy Ghost.[131] The table was thus a representation of both the Trinity and the cosmos in its substance, motion, and
coherence.

126. Ibid., p. 77.   127. Ibid., p. 94.   128. Ibid., p. 77.
129. See, for example, ibid., p. 79.   130. Ibid., p. 73.
131. Ibid., p. 74.

Two days later, Nalvage reappeared with another table. Dee noted that its division into ten or eleven sections distinguished it from the earlier mother-of-pearl table. It would have mattered very much to Dee from which table, and even from which section of the table, the characters were drawn, because their location would indicate whether they were related to the Father, the Son, or the Holy Ghost. By extension, this information would have related to the substance, action, and order of what the words designated. Even without any explicit link between the first and second tables, Nalvage's earlier remarks would have been on Dee's mind, and his synthetic skills would have been put to use forging connections between the tables.

Dee's efforts to distinguish between substance, action, and order were cast in a different light a few days later when Edward Kelly questioned the angels' use of both number and letters in their discussions of the divine language. Gabriel responded cabalistically: each letter's numerical designation, when combined into a word, signified the "member" or formula of that substance to which the entire word referred.[132] Every word, taken as a whole, represented the substance's "quiddity" or true essence. Letters and numbers thus provided a formulaic relationship between the divine language and the Book of Nature. A cabalistic exegesis of these formulas using the *gematria* method could help Dee to reorder the disordered and decaying world, as Gabriel confirmed when he explained that the letters on Nalvage's tables were "separated and in confusion," and had to be rearranged using their numerical formulas before they could be given to Dee.[133]

Numbers were equally important in the angels' instructions surrounding the thirty "Airs" and the angels of the twelve ancient tribes of Israel who governed them. After conveying a few of the Keys to open the Gates of Nature, the angels abruptly switched gears and began introducing Dee to what has been seen as a dramatically new system.[134] Though the Airs had not been mentioned before in the diaries, there was no radical demarcation between the Airs and the Governors, or the Airs and the Gates. Instead, there was such continuity between the two parts of the cabala of nature that the thirty Airs were also the final thirty Keys which opened the Gates of Nature.[135] The Airs did differ from other parts of the cabala of nature, however, in their more apocalyptic tone and purpose. Though the Governors were to share their knowledge of the Book of Nature with Dee, and the Gates were to give Dee the power to judge and ultimately restore nature through the recovery of natural "formu-

132. Ibid., p. 92.   133. Ibid., p. 64.
134. See, for example, Whitby, *AWS* I:153–154.   135. *TFR*, p. 145.

las," the Airs were destined to play a crucial part in the final reordering of humanity.

The thirty Airs were described as "bands" in the heavens distributed at twelve-degree intervals around the three-hundred-and-sixty-degree circumference of the earth. This arrangement provided Dee with a restored zodiac on which to base a reformation of astrology, as well as the names of the twelve angels who governed the thirty twelve-degree divisions of Airs in the heavens. These twelve angels were known for overseeing the twelve tribes of Israel, and their appearance in the midst of the vast majority of idiosyncratic angels who fill Dee's angel conversations makes them stand out in sharp relief.

When we take Dee's apocalyptic convictions into account, their position in the cabala of nature becomes even more consequential. In the Judeo-Christian tradition, the angels who watched over the twelve tribes of Israel played a major role in the growth of separate nations and the divisiveness and dispersal of human society.[136] The twelve tribes of Israel originated with the twelve sons of Jacob and their families and continued to serve as the organizational basis for Israelite society until the death of Solomon, when ten tribes seceded to form the Northern Kingdom. This kingdom later fell to the Assyrians and the Israelites were deported, assimilated into Assyrian society (as recounted in 2 Kings 17:6) and, ultimately, dispersed throughout the earth. Both Jeremiah 31:4 and Ezekiel 37:16 prophesied the return of the ten tribes, and, especially in Ezekiel, the return of the lost tribes was linked to the final days.[137]

The Thirty Airs in Dee's cabala of nature combined biblical ideas about the tribes of Israel with cabalistic ideas concerning the theurgic potential of secret names and apocalyptic ideas about the restitution of the Book of Nature. Each of the thirty Airs contained three divisions, and each Air was governed by an angel of one of the twelve tribes. A

136. The information for this and the subsequent paragraph was drawn from articles on the ten and twelve tribes of Israel contained in R. J. Zwi Werblowsky and Geoffrey Wigoder, eds., *The Encyclopedia of the Jewish Religion* (New York: Adama Books, 1986), p. 390.

137. In Ezekiel's vision, the twelve gates of the city of God (the apocalyptic Jerusalem) were inscribed with the names of all twelve tribes of Israel (Ezekiel 48:31–34). The names over the gates were to restore the tribes to order. Ezekiel's vision of the holy Jerusalem was repeated in Revelations 21:10–27, where the city was described in more detail. Revelations 21 was not the only place where the twelve tribes were mentioned, however. They were also a prominent feature in John the Divine's vision of the Last Judgment, when 144,000 people drawn from the twelve tribes were marked to save them from God's wrath (Revelations 7:1–8).

single angel could oversee several divisions within the heavens, and they were assisted by armies of other angels – a total of 491,577 angels, to be exact. The large number of angels was needed because each of the 90 celestial divisions corresponded to a discrete region of the terrestrial world and the people who lived there. It is quite likely that Dee understood the armies of angels as guardians of human souls – only 144,000 of which were to be saved, according to the Book of Revelations – that would help to sort the "marked" from the "unmarked" in the Last Judgment.

In keeping with other branches of the cabala of nature, Dee was given the seven-letter "true names" for the geographic regions supervised by the Airs, but was told that the thirty Airs had no names – they were known by their numbers.[138] Later, however, Dee was given three-letter names for the thirty Airs and was told that the names designated their "substance." Dee was dismayed by the discrepancy, which was never resolved.[139] Despite these contradictions, the numerological consistency of the seven-letter geographic names must have pleased the occult and mathematical facets of Dee's intellect, just as the information that the true name of Egypt was in fact "Occodon," and that Mesopotamia was known to God and the angels as "Valgars" must have pleased his geographical sensibilities. Dee's efforts to link his new system to his substantial geographical knowledge might have been too far-reaching, however, for Nalvage had to caution him that God did not practice the geography of Ptolemy or other terrestrial geographers.[140]

Not even Edward Kelly's "discovery" of some of the true names in the works of Agrippa, or his scryer's adamant refusal to have anything more to do with angels that had to look in books for their information, dampened Dee's enthusiasm for the Airs and their part of the cabala of nature. Dee simply expressed his delight with Kelly's bibliographic expertise. Always the humanistic scholar, Dee also had been searching in his library for more information on the geographic divisions of the earth in works by Mercator and Méla, but had met with no success.[141] If he was at all suspicious of Kelly or his motives in producing this damning evidence against the angels, he did not betray it. Nor was this incident, in fact, likely to have made Dee suspicious. After all, some shreds of "true" information should have survived the centuries.

Through the thirty Airs, Dee believed he was being given the means to restore the people of the earth to their ancient tribal associations. Despite the fact that the people of the twelve tribes were scattered all over the

138. *TFR*, p. 140.   139. Ibid., p. 209.   140. Ibid., p. 153.
141. Ibid., pp. 158–159.

globe, the angels were still responsible for them and knew where each tribe's descendants could be found. Dee's information regarding the twelve angels could be used to gather all people under the supervision of their rightful angel – even if they were (as in the case of the angel of the twelfth tribe) stretched out from Mauritania to the South Pole, to Chaldea, to the North Pole, and beyond.[142] On that long-awaited day when the city of God descended from heaven, the people of the earth would be able to enter the gates in an orderly fashion – and they would have Dee to thank for it. Although there is something humorous in the process by which Dee received the information concerning the thirty Airs – from the long days when names and places were given, to his hasty addition of the number of troops in angelic armies and Kelly's triumphant production of Agrippa – the ultimate purpose for this information would have imbued the proceedings with a momentous tone.

Dee's anticipation heightened when he received the next component of his cabala of nature, the Book of Enoch. Though this book has provided historians with a short, if inaccurate, rubric for Dee's entire angelic system, the materials associated with Enoch were not the only (or even the most central) feature of the cabala of nature.[143] The Book of Enoch was simply one of the materials Dee believed could bring him closer to a perfect understanding of the natural world. The angels first instructed Dee to make a book of forty-eight leaves.[144] One page from Enoch's book was not actually revealed to him, because its information was being witheld for God's use in the final days. The book was to contain both an "Invocation of the names of God" and invocations "of the angels by the names of God."[145] The angels told Dee that the information in the Book of Enoch would relate both to the heavens and to the future. Ontologically Dee was climbing higher and higher on the angelic ladder of wisdom and knowledge. Yet this description of the Book of Enoch contradicted subsequent revelations. At one point, Dee was instructed to seek answers to questions about "human affaires" in Enoch's book because it was "worldly."[146] If Dee tried to resolve this discrepancy between a supposedly heavenly book and a "worldly" book, we no longer have evidence of it, and his energies were instead diverted to desperate attempts to preserve his rapidly crumbling position in Prague.

142. Ibid., pp. 146–153.
143. See, for example, *The Enochian Evocation of Dr. John Dee*, ed. and trans. Geoffrey James (Gillette, NJ: Heptangle Books, 1988).
144. *TFR*, p. 159.   145. Ibid., p. 189.   146. Ibid., p. 394.

Several features of the Judeo-Christian tradition are relevant to the high place of Enoch in Dee's cabala of nature.[147] Enoch was a seventh-generation descendant of Adam, and he was also the seventh son of Noah. Thus, he fit numerologically into the cabala of nature, with its reliance on combinations of elemental and spiritual numbers. Enoch lived prior to the devastation of the Flood, and so was closer to the Adamic past as well. To mystics, however, the references to Enoch in Genesis that describe him as someone who "walked with God" and did not die but was "taken" by God (Genesis 5:21–24) were far more intriguing. These references suggest that Enoch (like Elias) knew about God, the angels, and the natural world in ways that other humans did not. Enoch's mystical connotations received more attention from Jewish authors, especially *Merkavah* mystics who believed that when Enoch was taken by God he was transformed into the archangel Metatron. Though this association was never fully endorsed in Talmudic scholarship, the angel Enoch/Metatron was revered as a teacher and a heavenly guide, and through these associations entered into the Western European magical tradition. Nor was Enoch ignored by more orthodox Christian commentators, who saw in his transcendence a prophetic precursor of Christ; and authors such as Tertullian and Irenaeus used Enoch to support the authenticity of Christ's physical resurrection.

Given these intriguing and powerful connections, it is unfortunate that the Book of Enoch is the least lucid part of the cabala of nature as it survives today. The angels' instructions are contradictory in several key places. Dee was distracted from the angel conversations by his attempts to acquire Rudolph II's support and his efforts to evade the papal authorities. Edward Kelly's attention was also absorbed by other matters, and he began to indicate that his allegiances were shifting from Dee and the angels to the emperor's alchemical experiments. In addition, we are missing significant parts of the angel diaries from the time during which the angels conveyed the Book of Enoch, and Dee might have been given other books to use with the book that were consumed in his furnace after the breakdown of relations with Pucci. We know that Dee had a

147. Information on Enoch for this and the subsequent paragraph was gathered from Steven D. Fraade's article in Mircea Eliade, ed., *The Encyclopedia of Religion*, vol. 5 (New York: Macmillan Publishing Company, 1987), pp. 116–118, and Gustav Davidson, *A Dictionary of Angels, Including the Fallen Angels* (New York: Macmillan, 1967), pp. 192–193. Further information on the book of Enoch can be found in Jósef T. Milik, ed., *The Books of Enoch* (Oxford: Oxford University Press, 1976) passim.

book containing "the Science of understanding of all the lower creatures of God," which was to be used with Enoch's book.[148] Earlier, Dee had been given four tables which he believed were "full of matter," including medicine, the knowledge of all elemental creatures, a complete knowledge of metals and stones (including their locations and uses), the knitting together of nature and its destruction, a knowledge of mechanical crafts, and instruction in formal, though not essential, alchemical transmutation.[149] These tables might have comprised the book entitled *The Mystery of Mysteries and the Holy of Holies*, which Dee confessed he did not then understand and which is no longer extant.[150]

The Book of Enoch represents a disappointing final note in the cabala of nature. Though the elements of a system can be discerned, a real understanding of how the system was intended to operate eludes the modern reader, as, one suspects, it eluded Dee. The scattered and fragmentary manuscripts of the angel diaries may yet be coaxed into yielding information, but Dee's cabala of nature remains an aspect of his thought that needs further exploration and explication. What is clear, however, is that Dee's belief in the utility of his cabala of nature was similar to the beliefs of other Christian cabalists, including Agrippa, who thought that cabala gave its practitioners a knowledge of the natural world and "the operative power to change the conditions of existence."[151]

148. *TFR*, p. 363.   149. Ibid., pp. 176–184.
150. Dee in Josten, p. 249.
151. Stephen A. McKnight, *Sacralizing the Secular: The Renaissance Origins of Modernity* (Baton Rouge and London: Louisiana State University Press, 1989), pp. 74–75.

6

Adam's Alchemy

The Medicine of God and the Restitution of Nature

> Behold, I show you a mystery;
> We shall not all sleep, but we shall all be changed,
> In a moment, in the twinkling of an eye, at the last trump: for
> the trumpet shall sound, and the dead shall be raised incorruptible,
> and we shall be changed.
>
> —1 Corinthians 15:51–52

Once he had attained mastery of the divine language and the cabala of nature, Dee believed that he was in a better position to discern and decipher corruptions in the Book of Nature. While this might have satisfied some natural philosophers, Dee conceived a much greater role for natural philosophy in the improvement of the human condition. The question facing him in the 1580s and 1590s was how to *practice* natural philosophy in a deteriorating world. This question was answered, and the final aspect of Dee's newly revealed natural philosophy took shape, when the angels delivered to him a form of alchemy that had not been practiced since the days of Adam.

Much of the rationale behind Dee's alchemy, both before and during the angel conversations, was linked to his belief in a "sickness" affecting the Book of Nature. Yet humanity was unable to diagnose or cure the Book of Nature without divine help, for alchemy itself had decayed and was unable to restore Nature's health. With the angels' assistance, Dee believed that he could restore the art of alchemy and then bring about a restitution of nature. The healing process would be based on his mastery of the "medicine of God," a divine substance that promised to redeem matter. The "medicine of God" was linked in important ways to Dee's other revealed science, the cabala of nature, and provided a unified base for the new natural philosophy.

When the apocalyptic and restorative features of the angel conversations are drawn together it is possible for us to understand why Dee believed that the conversations represented his highest achievements as a Christian natural philosopher. Dee was concerned with the consequences of the moral Fall and the subsequent afflictions of the Book of Nature, yet contemporary natural philosophy was unable to arrest, or even comprehend, the deterioration. Dee's new tools – including his showstone, the divine language, the cabala of nature, and other lessons learned in the angels' "celestial school" – would help him grasp the intricate workings of the Book of Nature and judge their decline. Alchemy was the last subject covered in the angels' lessons, and the great rewards Dee was promised once he mastered the practice of "Adam's alchemy" place the angel conversations even more firmly within the context of early modern natural philosophy.

The Promise and Practice of Alchemy

The role of alchemy in early modern natural philosophy, and in the thought of the period more generally, has been assessed by a number of scholars who offer evidence that Dee's belief in the healing, perfecting, and restorative powers of alchemy was not unique.[1] Alchemy's reputation as an art practiced by disreputable and desperate men demonstrates that the aspirations of natural philosophers like Dee were not always recognized by people outside alchemical circles. There are two reasons for the misperception. First, not all alchemists were philosophically inclined, as was Dee. Regular prosecutions of counterfeiters posing as alchemists fueled the popular belief that alchemists were fraudulent. Second, the obscure narratives used by philosophical alchemists in an

1. Two classic and reliable early studies are H. Stanley Redgrove, *Alchemy: Ancient and Modern* (New York: Barnes and Noble, 1922), and John Read, *Prelude to Chemistry: An Outline of Alchemy* (Cambridge, MA: MIT Press, 1966). More recently, William Shumaker has described aspects of alchemy in *The Occult Sciences in the Renaissance* (Los Angeles: University of California Press, 1972), pp. 161–198. Scholars who have studied particular figures or regional centers include R. J. W. Evans, *Rudolf II and His World* (Oxford: Oxford University Press, 1973); Betty Jo Teeter Dobbs, *The Foundations of Newton's Alchemy* (Cambridge: Cambridge University Press, 1975), Bruce T. Moran, *The Alchemical World of the German Court: Occult Philosophy and Chemical Medicine in the Circle of Moritz of Hessen (1572–1632)*, Sudhoffs Archiv Zeitschrift für Wissenschaftsgeschichte, Beiheft 29 (Stuttgart: Franz Steiner Verlag, 1991); William R. Newman, *Gehennical Fire: The Lives of George Starkey, an American Alchemist in the Scientific Revolution* (Cambridge, MA: Harvard University Press, 1994); and Pamela H. Smith, *The Business of Alchemy: Science and Culture in the Holy Roman Empire* (Princeton, NJ: Princeton University Press, 1994).

effort to protect their secrets made material transformation seem even more mysterious than it actually was.

Some authors explicitly discussed practical applications of the alchemical art, including the Siennese metallurgist Vannoccio Biringuccio (1480–1539?). Biringuccio's *Pirotechnia*, published posthumously in 1540, includes metallurgical information as well as a discussion of more obscure alchemical topics.[2] Such a work can help the modern reader to locate the angelic alchemy in a firmer, less allegorical context than Dee would have used. Biringuccio would not have been surprised at Dee's desire to learn about alchemy from angelic schoolmasters, for example. Biringuccio believed, after examining the ideas of the alchemists, "that unless someone should find some angelic spirit as patron or should operate through his own divinity," the art was doomed to failure.[3] He also felt that alchemy could never truly replicate natural processes, despite assertions that it could hasten the growth and perfection of the natural world. Nature, Biringuccio wrote, operated "in things from within" and caused all things to grow and mature according to a gradual, divine time frame.[4] Alchemy, on the other hand, was an art "weak in comparison" that trailed after "Nature in an effort to imitate her," attempting to bring about change quickly and unnaturally.[5] Hastening nature, Biringuccio pointed out, was not feasible, as was evident from the vast resources squandered by princes on alchemical schemes.[6]

Despite Biringuccio's skepticism, he could not banish alchemy from the respected enclaves of learning altogether, in part because of its laudable objectives in the hands of truly adept, religious practitioners. Biringuccio distinguished between the "just, holy, and good" alchemy practiced by natural philosophers such as Dee, and the "sophistic, violent, and unnatural" alchemy practiced by charlatans.[7] Those who practiced the "holy" type of alchemy believed they were "true physicians of mineral bodies, purging them of superfluities and assisting them by augmenting their virtue and freeing them from defects."[8] Though Biringuc-

---

2. For a modern edition of this work, see Vannoccio Biringuccio, *The Pirotechnia of Vannoccio Biringuccio*, trans. and ed. Cyril Stanley Smith and Martha Teach Gnudi (New York: Dover Publications, 1990). Dee owned two copies of the work: the original Italian edition of 1540 (Roberts and Watson, #677/B108), which he took with him to the Continent in 1583, and a French translation of 1556 (Roberts and Watson, #459), which remained in England. Neither has been recovered.

3. Biringuccio, *Pirotechnia*, p. 35.   4. Ibid., p. 37.

5. Ibid., pp. 37–38.   6. Ibid., p. 36.   7. Ibid., pp. 336–337.

8. Ibid., p. 336.

cio confessed that he did not know whether any alchemist had been able to cure the natural world by alchemical means, he conceded that the "sweetness offered by the hope of one day possessing the rich goal this art promises . . . is surely a fine occupation."[9]

Alchemy, like other branches of early modern natural philosophy, was vitalistic. Its symbolism suggested connections between the alchemist's work in the laboratory and the divine process of cosmic creation, decay, and restitution as presented in the Book of Scripture. Alchemy posited masculine and feminine principles, for instance, which combined to give birth to a "child"; metals underwent "generation," "digestion," and "putrefaction"; and the work itself took place in a closed vessel that alchemists called a "womb."[10] In an analogue to biblical accounts of cosmic history, the alchemist subjected his metals to a process of transformation that progressed from chaos, to differentiation, decay, and death, until, finally, the metals were "resurrected" and became the philosophers' stone – a marvelous substance that could cure all sickness, render the alchemist immortal, and perfect nature.[11] Thus it was possible for alchemists to draw a parallel between their creation of the philosophers' stone and the resurrection of Christ's body.

This parallel was based on a pervasive belief that death and decay were necessary if rebirth and restitution were to take place. As the German alchemist and physician Paracelsus (1493–1541) explained in his *Die 9 Bücher de Natura rerum* (1537):

> Decay is the midwife of very great things! It causes many things to rot, that a noble fruit may be born; for it is the reversal, the death and destruction of the original essence of natural things. It brings about the birth and rebirth of forms a thousand times improved. . . . And this is the highest and greatest *mysterium* of God, the deepest mystery and miracle that He has revealed to mortal man.[12]

Although the decay of the Book of Nature made it difficult to read its contents with any degree of accuracy, decay itself could still be a form of revelation to natural philosophers, because God was showing humanity the way into the future through gradual deterioration. The culmina-

9. Ibid., p. 337.
10. Betty Jo Teeter Dobbs, *Alchemical Death and Resurrection: The Significance of Alchemy in the Age of Newton* (Washington, DC: Smithsonian Institution Libraries, 1990), p. 4.
11. Ibid., p. 15.
12. Quoted in Paracelsus, *Selected Writings*, ed. Jolande Jacobi, Bollengen Series, vol. 28 (Princeton, NJ: Princeton University Press, 1988), p. 144. Dee's copies of *Die 9 Bücher de Natura rerum* (Roberts and Watson, #1485 and #2268) have not been recovered.

tion of this process was restitution, followed by the repetition of the cycle. Philosophical alchemists like Dee were in a unique position to help society follow the path toward redemption because of their practical knowledge of the cycle of decay, death, and ultimate rebirth.

The goal of every philosophical alchemist was the philosophers' stone, a powerful substance capable of perfecting other substances with which it came into contact. Analogies between the perfecting power of the philosophers' stone and the healing power of medicines were not lost on alchemical practitioners. Paracelsus was particularly influential, because he elaborated and modified traditional alchemical ideas to link alchemy (healing the imperfections in nature) with medicine (healing bodily sicknesses and imperfections).[13] In Paracelsus, alchemy served as the foundation for medicine, as well as the crucial discipline needed for inquiry into the Book of Nature. He argued that alchemy brought the natural philosopher closer to the secrets of nature and God because it was based on redemptive principles and Christian theological concepts.[14]

The ideas of Paracelsus and other alchemical thinkers of antiquity, the Middle Ages, and the early modern period were well known to Dee. Alchemy was one of Dee's lifelong passions; the spiritual and material quest for the philosophers' stone was as appealing to him as communicating with angels. The works of Paracelsus occupied a significant place in Dee's library, and though few of his personal copies have been recovered, the sheer number of volumes in the collection indicate that he was familiar with Paracelsus's ideas. He did not embrace the philosophies of Paracelsus in their entirety, however: rather than adopting his three-principle (salt/mercury/sulphur) theory of alchemy, Dee's *Monas hieroglyphica* continued to use the two-principle (mercury/sulphur) theory derived from Arabic alchemy.[15] Still, a combination of traditional and Paracelsian ideas provide the background for the angels' restored alchemical art.

A better understanding of Dee's alchemical beliefs can be gleaned from the *Monas hieroglyphica*, especially theorems XVIII and XIX. Both theorems connected the symbol of the *monas* to the process of alchemical transmutation. In alchemy, according to Dee, heat was applied to ele-

---

13. The most comprehensive study of Paracelsus can be found in Walter Pagel, *Paracelsus* (New York: S. Karger, 1958). For additional interpretations, see Charles Webster, *From Paracelsus to Newton: Magic and the Making of Modern Science* (Cambridge: Cambridge University Press, 1982), and Allen G. Debus, *The English Paracelsians (1965)* (New York: Franklin Watts, 1966).

14. Debus, *English Paracelsians*, p. 41.   15. Clulee, pp. 96–98.

ments in a closed vessel, but the heat was mediated and made gentle by the protective use of dung. To explain the components of the closed vessel and their transformation, Dee first used a representation of an egg, a common alchemical symbol.[16] Dee arranged the elements of the transformation along orbital lines in the egg according to the corresponding placement of their ruling planets. He delineated the egg further into a central "yolk" made up of Sol, Mars, and Venus (the metals gold, iron, and copper) and a "white" made up of Jupiter, Saturn, Luna, and Mercury (the metals tin, lead, silver, and quicksilver). Dee never clarified how the "chalk" or shell of the egg was comprised, though it did play a role in transmutation. Material transformation took place when the "yolk" enveloped the "white" through a process of "rotation." In the *Monas hieroglyphica* the process was symbolized by a spiral derived from Arabic alchemy, where "turning the screw" was used to describe transmutation.[17]

Dee linked the seven rotations or revolutions in the alchemical process of transmutation to the ultimate goal of philosophical alchemy: the divinization of the alchemist. While "vulgar" alchemy was concerned only with the material transformation of base metals into silver and then gold, "philosophical" alchemy attempted to transform the alchemist's soul from human to divine. This, too, was symbolized in the *monas*. Dee explained that the terrestrial body in the center of the *monas*, when activated by a divine force, would be united in a perpetual marriage to a generative influence. When the marriage took place under appropriate solar and lunar influences, then the terrestrial center of the *monas* could no longer be "fed or watered . . . until the fourth, great, and truly metaphysical revolution was completed." After this material "advance" occurred, the alchemist "who fed [the monad] will first himself go away into a metamorphosis and will afterwards very rarely be held by mortal eye." Dee was probably referring to a particular step in the seven revolutions of the alchemical process when he mentioned the "fourth, great, and truly metaphysical revolution." Ontologically, the fourth revolution would have been the mediating step between lower and higher levels of the alchemical process and would occupy the same intermediary position in alchemy as angels and mathematics did in the cosmos.

These references in the *Monas hieroglyphica* demonstrate that Dee understood the traditional division of alchemy into two different branches of inquiry. Practical alchemy was primarily interested in turning lead or other base metals into gold. Alchemical texts varied regarding

16. H. J. Sheppard, "Egg Symbolism in Alchemy," *Ambix* 6 (1958): 140–148.
17. For the principles of the Arabic mercury-sulphur theory and its relationship to Dee's *monas*, see Clulee, pp. 97–100.

how this transformation from base matter was to take place, but in its most rudimentary form it involved the purification of various materials, the application of heat while the materials were in an enclosed vessel to engender color changes, increasing the strength of the mixture through multiplication, and finally projecting the resulting mixture onto other matter to cause transmutation.[18] Spiritual alchemy was a more complex and philosophical process wherein the ascent of the matter toward the philosophers' stone was a material reflection of the spiritual and intellectual ascent of the alchemist toward God. As in many occult sciences, successful alchemists were thought to have been divinely selected to achieve their goal because of their inner purity and strength of purpose. Some equated the soul of the alchemical philosopher with the material to be transmuted: when the material in the alchemical vase reached a state of quintessence, then the soul of the philosopher would reach a state of spiritual enlightenment and power.

Adam's Alchemy: Restoring the Alchemical Art

The angels' alchemy did not correspond exactly to the philosophical alchemy practiced by Dee and his contemporaries. Like the cabala of nature, the angelic alchemy was interesting to Dee precisely because it did *not* correspond exactly with what he knew, and thus could be seen as an improvement in the state of learning and knowledge. As in the case of the cabala of nature, Dee faced each discrepancy between angelic and traditional alchemy with a firm belief that he was receiving Adam's "true" alchemy. Alchemical motifs, ideas, and parables in the angel conversations, therefore, are analogous to, but not strictly derivative from, the alchemy Dee had been practicing.

One of the benefits of practicing "Adam's alchemy" would have been its resolution of the early modern alchemical competition between the ancient sulfur-mercury theory of elemental transformation and the new three-principle theory articulated by Paracelsus, which added salt.[19] Dee purchased many books by Paracelsus – well over a hundred – and Paracelsian ideas, which would not be widely accepted in England until the seventeenth century, were familiar in the university culture of Louvain when Dee was there.[20] So why did Paracelsus not persuade Dee on

18. See Read, *Prelude to Chemistry*, pp. 118–163.
19. Ibid., pp. 25–27; Debus, *English Paracelsians*, p. 27.
20. Thorndike, V:329. For Paracelsus's adoption by English natural philosophers, see Debus, *English Paracelsians*, especially pp. 55–57, although his belief that "It was not until 1585 that any Englishman took notice of the

this pivotal aspect of alchemical theory? Though Paracelsus's influence on Dee has been insufficiently studied, the list of titles he owned suggests that it was manifested in surprising ways, and that his influence may not have been primarily alchemical.[21] Five copies of two of Paracelsus's early works on medicine, the *Paragranum* and the *Opus paramirum*, attest to Dee's interest in Paracelsian "universal" medicine despite the fact that Dee was not a practicing physician.[22] Dee also owned later works by Paracelsus, including ten treatises on surgery[23] and seven copies of his comprehensive survey of occult philosophy, the *Archidoxa*, which is reminiscent of works by both Agrippa and Johann Trithemius.[24]

It is likely that the debates between Paracelsian and Arabic alchemy

comprehensive Paracelsian theories" should be qualified in light of Dee's collection and contemporary (though negative) Elizabethan reactions.

21. Clulee makes only one reference to Paracelsus (p. 141). French makes more frequent allusions to the chemical physician, but they are not extensive. See Peter French, *John Dee: The World of an Elizabethan Magus* (New York: Routledge and Kegan Paul, 1972), pp. 1, 37, 52, 60–61, 76, 127–128, and 136. In addition, Dee is not featured in Allen Debus' *English Paracelsians*, but only mentioned on pp. 101 and 131.

22. Paracelsus, *Paragranum* (Frankfurt, 1565), Roberts and Watson, #1491; *Paramirum* (1562), Roberts and Watson, #1467 and #2249; *Paramirum* (Frankfurt, 1565), Roberts and Watson, #1490; *Paramirum* (Augsburg, 1575), Roberts and Watson, #1507; and *Paramirum* (Basel, 1570), Roberts and Watson, #2229.

23. Paracelsus, *Chirurgia magna* (Frankfurt, 1566), Roberts and Watson, #65; *Cheirurgia magna* (Basel, 1573), Roberts and Watson, #70; *Cheirurgiae libri tres* (Frankfurt, 1562); *Chirurgia magna* (Augsburg, 1566), Roberts and Watson, #1479; *Chirurgiae magnae partes tres* (Frankfurt, 1562), Roberts and Watson, #1480; *Chirurgiae Magnae tractatus tres* (Antwerp, 1556), Roberts and Watson, #1500; *Cheirurgia minor* (Basel, 1570), Roberts and Watson, #1503; *Cheirurgia magna et minor* (Basel, 1573), Roberts and Watson, #2220; *Cheirurgia vulnerum* (Basel, 1573), Roberts and Watson, #2230; *Chirurgia minor* (Basel, 1573), Roberts and Watson, #2231.

24. Paracelsus actually studied under Trithemius, as did Agrippa. See Paracelsus, *Archidoxa* (Basel, 1572), Roberts and Watson, #1474; *Archidoxa* (Munich, 1570), Roberts and Watson, #1475; *Archidoxa* (Augsburg, 1574), Roberts and Watson, #1492; *Archidoxa* (Augsburg, 1570), Roberts and Watson, #1494; *Archidoxa* (Basel, 1570), Roberts and Watson, #1502; *Archidoxa* (Basel, 1570), Roberts and Watson, #2227; and *Archidoxa* (Augsburg, 1570), Roberts and Watson, #2263. The *Archidoxa* is available in a reprint edition of the 1656 translation by Robert Turner. See Paracelsus, *The Archidoxes of Magic* (1656), trans. R[obert] Turner (New York: Samuel Weiser, 1975).

did not trouble Dee, because the angels were revealing new, uncorrupt information about alchemical materials and processes.[25] Dee still adhered to the two-principle theory of transmutation in 1564 when the *Monas hieroglyphica* was published, and the angels' use of the two-principle theory ultimately resolved the question as far as Dee was concerned. Dee discovered, for example, that "Dlafod" was the real name for sulfur, and "Audcal" the true name for mercury. The angels also shared the names of other alchemical materials, as well as instructions about the mysterious processes leading to "Dar," or the philosophers' stone. Dee needed to use the cabala of nature to reach the philosophers' stone, so that he could "judge, not as the world doth, but perfectly of all things contained within the Compasse of Nature, and of all things which are subject to an end."[26] Then, Dee would be able to apply the *medicina dei* or *medicina vera* – the medicine of God or true medicine – to the Book of Nature to help restore and heal it. Gabriel discussed the *medicina dei* with Dee, where it was described as "the true, and perfect science of the natural combination, and proportion of known parts."[27] In order to heal any sickness within nature, Dee had to know both "celestial radiation," which was the cause of natural combination, and "elemental vigor," which was the cause of natural proportion.[28] Dee was impressed, and found the lesson on the medicine of God "very apt."

The angels linked the *medicina dei* to the cabala of nature and the divine language, telling Dee that the power in the angels' words was akin to the alchemical properties of the medicine. Dee found this notion appealing, perhaps because he could relate it to Paracelsus's belief that all natural philosophers must base their exploration of the Book of Nature on the "true foundation" provided by the cabala.[29] After Paracelsus's death, "R. Bostocke" endorsed this idea in a work entitled the *Difference betwene the auncient Phisicke . . . and the latter Phisicke* (1585). Bostocke wrote: "The true auncient phisicke which consisteth in the searching out of the secretes of Nature . . . collected out of the Mathematicall and supernaturall precepts . . . is part of Cabala. . . ."[30] This symbiotic relationship between alchemy and cabala is evident in Dee's angel conversations, where both were integral components of a newly reformed and revealed natural philosophy.

The physics and metaphysics of light were also important to the *med-*

25. *TFR*, pp. 387–388. 26. Ibid., p. 77. 27. Ibid., p. 251.

28. Ibid. 29. Paracelsus, *Selected Writings*, pp. 133–134.

30. R[obert?] Bostocke, *The Difference betwene the auncient Phisicke . . . and the latter Phisicke* (London, 1585), sig. Biᵛ discussed in Debus, *English Paracelsians*, p. 61.

icina dei, because only a natural philosopher proficient in "celestial radiation" could manufacture the substance. Dee's fascination with Giovanni Pantheus's *Voarchadumia contra alchimiam* provides insight into the unification of alchemy and optics, as do Dee's published works.[31] In the *Voarchadumia,* Pantheus cabalistically equated the letter O with both "lux major" and "lux minor"; the letter A with "materia prima artis," or elemental first matter; and the letter S with fire or air, which were "lux minor."[32] Depending on the combination of these elements, and the desired transformation, Pantheus assigned each letter a numerical weight, or proportion. In the *Propaedeumata aphoristica* Dee modified Pantheus's ideas and argued that each of the four elements was comprised of different proportions of elemental matter, lux major, and lux minor (which was itself composed of a mixture of air and fire).[33]

Light played a significant role in Dee's alchemical philosophy, and thus alchemy became another science susceptible to opticomathematical interpretation. When Dee referred to alchemy as "inferior astronomy," he was not simply employing a common trope but expressing his belief in the mathematical connections that might exist between celestial light and elemental matter. Robert Grosseteste's metaphysics of light was influential in the formation of Dee's interest in optics, and he might have influenced Dee's beliefs in alchemy and light as well. Like Dee, Grosseteste believed that light governed the interaction of elements and elemental change and that both processes were regulated and directed with the mediation of the celestial bodies.[34]

---

31. Dee may have received further insights into Pantheus's theories from Giovanni Baptista Agnelli, a Venetian alchemist living in London. Roberts and Watson persuasively argue that Agnelli gave Pantheus's work to Dee based on Dee's inscription "ex dono magister Johannes Baptistae Danieli" on the title page. Is is not inconceivable that this was Dee's erratic (and largely phonetic) latinization of Agnelli's name. See title page, Joannes Augustinus Pantheus, *Voarchadumia contra alchimiam: ars distincta ab archimia, & sophia: cum Additionibus: Proportionibus: Numeris: & Figuris* (Venice, 1530). Dee's annotated copy is now in the British Library, Roberts and Watson, #D16. Agnelli is a fascinating and neglect figure in the history of English alchemy, but he was well known in the period and participated in the assays and trials of Martin Frobisher's mysterious gold in the late 1570s. Dee had a copy of Agnelli's book on alchemy, the *Apocalypsis spiritus secreti* (London, 1566), in his library; Roberts and Watson, #1449. Dee took Agnelli's work with him to the Continent in 1583; it has not been recovered.

32. Pantheus, *Voarchadumia,* pp. 17ᵛ-18ʳ.　33. Dee, *PA,* pp. 129–131.

34. James McEvoy, "The Metaphysics of Light in the Middle Ages," *Philosophical Studies (Dublin)* 26 (1979): 135.

Enacting Adam's Alchemy

When the angels instructed Dee in Adam's alchemy, it was always a lively, theatrical exchange. The angels' alchemical ideas were ideally suited to visual and oral methods of transmission. Color and figural symbolism played a prominent role, for example, for alchemical materials were thought to alter in color from black to white, from white to green or yellow, and finally from green or yellow to red. While a figure who appeared in the showstone in red garments bearing a sword might in traditional symbolic terms signify war or the god Mars, in alchemical terms such a figure would convey additional meanings relating to stages in the alchemical process, metals, and planetary influences. Symbolic representations of the alchemical process known as *emblemata* were popular throughout the early modern period and were first issued in printed form in the seventeenth century.[35] Materials and steps of the alchemical process were personified in contemporary texts, as when masculine and feminine figures representing Sophic Sulphur and Sophic Mercury joined in a chemical wedding. The philosophers' stone resulted from the union, variously personified as an androgyne or "royal child." In one of Dee's conversations, a boy wearing a crown transformed a woman into a man, and then the boy wed the androgyne to a queen, just as if an emblem book had come to life within his showstone.[36] This shared matrix of symbols rests on what William Ashworth calls the "emblematic world view," which he defines as "the belief that every kind of thing in the cosmos has myriad hidden meanings and that knowledge consists of an attempt to comprehend as many of these as possible."[37]

Unlike two-dimensional emblem books, the allegorical parables in the angel conversations were capable of movement and action. This unique

---

35. See, specifically, Gerard Heym, "Some Alchemical Picture Books," *Ambix* 1 (1937): 69–75, and Stanislas Klossowski de Rola, *The Golden Game: Alchemical Engravings of the 17th Century* (London: Thames and Hudson, 1988), passim. Mario Praz, *Studies in Seventeenth-Century Imagery* (Rome: Edizioni di storia e letteratura, 1975), also contains valuable information on the place of alchemical *emblemata* in the larger emblematic tradition.

36. *TFR*, p. 16.

37. William B. Ashworth, "Natural History and the Emblematic World View," in *Reappraisals of the Scientific Revolution*, ed. David Lindberg and Robert S. Westman (Cambridge: Cambridge University Press, 1990), p. 312. For another analysis of the use of images in the angel conversations, See Deborah E. Harkness, "Shows in the Showstone," passim, which was an early version of some material in this chapter.

feature of Dee's angel conversations enabled alchemical ideas to be mediated through a language of signs, colors, symbols, and gestures suited
to the didactic purposes of the celestial school. The angelic schoolmasters
spun series of images and actions into an alchemical narrative to explain
complex alchemical processes. Often these lessons included additional
instruction in the divine language, and Dee was given the "true" names
for stages and materials of the alchemical art. The parables also addressed the immediate apocalyptic significance of Dee's alchemical practices.

One of the richest didactic parables was shown to Dee during the
afternoon of 16 November 1582. John Dee and Edward Kelly were
together in Dee's private study in Mortlake. They had not waited long
when a male angel appeared in the stone dressed in a long, purple robe
with a triple crown on his head.[38] The angel brandished a red rod, and
the earth below him shook. Seven other angels approached, holding a
seven-pointed copper star. One of the seven spoke to Dee, saying, "I am
he which have powre to alter the corruption of NATURE. with [sic] my
seal, I seale her and *she [nature] is becom[m]e perfect*. I prevayle *in
Metalls*: in the *knowledge of them*." Another angel came forward and
told Dee, "I am *Prince of the Seas*: My powre is uppon the waters. I
drowned Pharao . . . My name was known to Moyses. I lived in Israel.
Beholde *the tyme of Gods visitation*." After prophecies about the destructive power of the sea, the angel opened his robe, revealing feathers
and a golden girdle. At this point, a "black cloth was drawn" inside the
stone, which Dee noted "is now appointed to be our token . . . that we
must leave of[f] for that instant."

After a break, Dee and Kelly resumed their conversation. The crowned
angel appeared, as did three angels still holding the copper star. One
stepped forward, saying, "*My powre is in Erth: and I kepe the Bodies of
the Dead*. Theyr members are in my bokes. I have the key of Dissolution.
. . . *Behold, the bowels of the earth are at my opening*." Dee, who was
feeling some financial pressure at the time, asked the angel for help in
finding hidden treasure so that he could pay his debts. The angel of the
earth chastised Dee, replying that the treasures hoarded in the earth were
reserved for the destruction of the Antichrist. Instead, the angel was
giving Dee the power of his "seal," which he would be able to use to
govern the earth and unlock its secrets. Another angel stepped forward,
who was "*the life and breath of all things in Living Creatures*." This
angel made birds, dragons, and other creatures appear in the stone, and
said, "*the Living, The ende, and beginning of these things*, are known

38. Dee, *AWS* II:147–157; all emphases are Dee's.

unto me: and by sufferance *I do dispose them untyll my Violl be run.*"
Then the angel held up a vial containing five or six spoonfuls of oil.
When the sixth angel pulled open his red clothes, fire issued forth from
his sides, which Dee reported "skarsly of mans eye can be beholden."

Finally, the seventh angel spoke to Dee and Kelly, saying "The *powres
under my subjection, are Invisible* . . . I will teache the[e] *Names without
Numbers. The Creatures subject unto me shalbe known unto you.* Beware of *wavering.* Blot *out suspi*tion of us for we are Gods Creatures,
that have rayned, do rayne, & shall rayne for ever. *All our Mysteries
shalbe known unto you.*" The conversation ended with some final words
from the first angel: "Behold, these things, and theyr mysteries shalbe
known unto you, reserving the Secrets of him which raigneth for ever.
. . . Whose name is Great for ever. . . . Open your eyes, and you shall see
from the *Highest to the Lowest.*" Once again the black curtain signaled
the parable's end.

This parable emphasized ideas that would have resonated with Dee's
natural philosophical interests and intentions. The first angel, whose
knowledge of metals permitted alterations in nature, had obvious links
to both his alchemical and his apocalyptic interests. The second angel
was of great antiquity and had advised Moses, linking Dee to the wonders of the past. Dissolution and decay were covered by the third angel,
while the fourth knew about the ending and beginning of all living
creatures. Promises of the knowledge of Adam, the Book of Nature, the
past, and the future were all contained in a single angel conversation. In
addition, the symbolism was provocative without being too specific,
which permitted Dee and his associates to discuss the "lesson" at length
and spin out its possible permutations.

One potential meaning of the parable that no one would have missed
was apocalyptic, for in this conversation apocalypticism and alchemy
collapse into a single narrative designed to convey the message that Dee
and Kelly could be instrumental in the world's progress toward the Final
Days. The first angel who spoke with Dee had the "power to alter the
corruption of nature" and would perfect nature, while the second angel
told Dee to expect the time of "God's visitation." The angels had been
divinely instructed to share their powers with Dee – and thus he, too,
could master the natural world.

Similarly alchemical messages were delivered repeatedly during the
course of Dee and Kelly's five-year association. Another conversation on
4 May 1582 was infused with an assortment of speeches, gestures,
special effects, and symbolism to deliver a similar message. On this
particular day Edward Kelly was professing doubts about their exercises
and refusing to begin the prayers that served as a preamble to each

attempt to contact the angels. Dee, undeterred by Kelly's doubts, went into his oratory "and called unto God, *for his divine help* for the understanding *of his laws and vertues . . . which he hath established in and amongst his Creatures for the benefyt of mankinde.*"[39] Soon Kelly reported seeing two angels, Michael and Uriel, who joined them in prayer.

Suddenly, seven bundles appeared "from heavenward." While Michael collected the bundles, Uriel placed "a thing like a superaltare" on the table "and with a thing like a Senser" made "perfume at the fowre corners of the Table." The censer sank through the middle of the table, and Uriel took the seven bundles from Michael and placed them on the superaltar. A "Glorious man" unwrapped the bundles, revealing animals. The "Glorious man" took a bird, "as byg as a sparrow," and the bird began to increase in size, seeming "to be as great as a swanne: very beutifull [*sic*]: but *of many cullours.*" Another bird went through the same metamorphosis, and the "glorious man" joined their wings together. Kelly reported that "All is suddenly dark, and nothing [is] to be seen." Then he heard a voice like the angel Michael's saying, "It was a byrd, and it is a byrd, absent their [*sic*] is nothing but Quantitie. Beleve. *The world is of Necessitie: His Necessity is governed by supernaturall Wisdome.* Necessarily you fall: and of Necessitie you shall rise again."

The showstone brightened and Michael and Uriel returned to show Dee and Kelly the remainder of the parable. The two birds reappeared, and grew until they were "*as big as mowntaynes.*" They flew toward heaven, their wings touching the sky. One bird took stars into his beak which the other bird replaced in the heavens. The birds then flew over cities and towns, striking down bishops, princes, and kings with their wings as they passed. The "Simple" folk – beggars, the lame, children, old men and women – were left standing, untouched by the birds.

At this point, the imagery took a different turn. The two birds lifted the bodies of three dead men and a child, all wearing crowns, from the ground. The four began to quicken from their death, and then rose up and departed. The birds rested upon a hill, where the first bird "gryped the erth mightily and there appeared diverse Metalls." The birds tossed an old man's head until it broke open, spilling out "a stone, rownd, of the bignes of a Tennez ball of 4 cullours, White, black, red, and greene." The birds nibbled at the stone, and were transformed into crowned men with golden teeth, hands, feet, tongues, eyes, and ears. The men carried "Sachels . . . full of gold," which they "seemed to sow . . . as corne, going or stepping forward, like Seedmen." Here the parable ended, and Michael informed them that he had shown Dee and Kelly how they

39. Ibid., pp. 130–138.

would be joined, by whom, to what intent and purpose, what they are, what they were, and what they should be.

Decoding the parable requires that it be divided into two symbolic cycles: the first extending from Uriel's actions at the altar until the kings and bishops were struck down; and the second extending from the revivification of the dead to the sowing of gold in the earth. The first cycle was consonant with traditional symbols, while the second relied heavily on alchemical imagery. At the beginning of the first cycle of images, Uriel's use of the censer to consecrate the altar bears a striking similarity to Revelations 8:3–6, where the consecration of an altar serves as a link between the opening of the seventh seal, which brings a profound silence to the world, and the sounding of the first trumpet of the war between good and evil. Kelly and Dee were cast into the parable as birds transformed from sparrows – the most humble of birds – into conjoined swans. Generally, birds symbolized the human spirit; throughout the Middle Ages the swan symbolized the soul's journey toward salvation through its ascension to the kingdom of heaven.[40]

Spiritual transformation and salvation were mimetically underscored by the words of the angel Michael, who said, "Necessarily you fall: and of necessity you shall rise again." That spiritual redemption would lead to mastery over the natural world was dramatized when the two birds reached into the heavens and reordered the stars. The reordering of the stars had apocalyptic overtones, and any changes in the heavens were a sign of great importance in the sixteenth century. The significance of the new heavenly order might have been revealed in the birds' next action, when they leveled the mighty with their wings, leaving the weak standing. In Revelations 6:15–17 a similar event serves as a precursor to the apocalyptic days of wrath:

> And the kings of the earth, and the great men, and the rich men, and the chief captains, and the mighty men, and every bondman, and every free man, hid themselves in the dens and in the rocks of the mountains; And said to the mountains and rocks, Fall on us, and hide us from the face of him that sitteth on the throne, and from the wrath of the Lamb; For the great day of his wrath is come; and who shall be able to stand?

The second cycle of images, though it has apocalyptic significance, was drawn primarily from the alchemical tradition. The cycle begins with the birds' revivification of the dead. Death followed by rebirth was a common motif in alchemical literature, and alchemical theory posited

40. Louis Charbonneau-Lassay, *The Bestiary of Christ*, trans. and abridged by D. M. Dooling (New York: Parabola Books, 1991), pp. 251–252.

that nothing could "be multiplied or propagated without decomposition."[41] A more explicit link to alchemy occurred when the earth opened to reveal the hidden world of metals, followed by the appearance of the philosophers' stone – the multicolored stone the size of a tennis ball – which the birds ingested. The divine food of the philosophers' stone transformed the birds (symbolizing Dee and Kelly) into men of perfect gold, thus linking the goal of the alchemical process with the redemption of their souls. Their perfection would enable them to sow the seeds of rebirth in the Book of Nature, as demonstrated when the two men began to sow their bags of gold in the earth.[42] These images were common in alchemical texts, and in Mylius's *Philosophia reformata* (1622) an engraving appeared which included, not only the revivification of the dead, but also the philosopher sowing gold and an angel blowing a trumpet signaling the Apocalypse, when the dead shall rise and be damned or redeemed (Figure 7).

Whereas the parables of 4 May 1582 and 16 November 1582 are replete with alchemical ideas, later alchemical parables contain a bewildering assortment of symbolic images that cannot always be made to yield a coherent alchemical narrative. In some cases, neither the color symbolism nor the personifications of elemental materials adhere to early modern alchemical theory: the colors appear throughout the actions in an incorrect order, or the alchemical characters appear in unusual sequences and combinations. Whether the jumble of images and ideas reflects alchemical developments invented by Dee and Kelly, or whether Dee and Kelly went on to practice the alchemical processes suggested by the angels' parables, is not clear.

Four angel conversations that took place between 1584 and 1585, when Dee and Kelly were in Prague seeking an audience with Rudolf II, are illustrative.[43] All contain allusions to birds, rivers, towers or castles, gardens, fire, and flowers. While we might be tempted to quickly pass over these images, they are common alchemical symbols.[44] Birds, for

---

41. Basil Valentine, quoted in Read, *Prelude to Chemistry*, p. 202.
42. For the significance of the vegetation of metals, see Dobbs, *Alchemical Death and Resurrection*, p. 10 and Read, pp. 94–95.
43. See conversations for 30 April 1584 (*TFR*, pp. 111–114); 20 June 1584 (*TFR*, pp. 168–171); 20 August 1584 (*TFR*, pp. 219–222); 14 January 1585 (Dee, *TFR*, pp. 355–361).
44. For further insights into the rich symbolic possibilities of alchemy, see Lyndy Abraham's *Dictionary of Alchemical Imagery* (Cambridge: Cambridge University Press, 1998). The information in this paragraph and in subsequent pages is drawn from Abraham's study.

Figure 7. An angel blows a trumpet to awaken the dead, while an alchemist sows seeds of gold, in this illustration of "Fermentatio," from Mylius's *Philosophia reformata*. Reproduced by permission of the University of Wisconsin, Madison Library Department of Special Collections.

example, indicated the combination of masculine and feminine principles into a "volatile spirit" – a crucial stage in alchemical transformation. Rivers or streams commonly denoted the "mercurial water" that transformed one element into another. The alchemical furnace was often depicted as a tower or castle in alchemical treatises, and some authorities went so far as to point out the advisability of constructing an alchemical furnace that was not only functional but visually pleasant.[45] Entire gardens could also symbolize alchemical equipment, in this case the closed vessel in which material transformation took place. Alchemists often protected their unique designs for closed vessels, referring to them in their treatises as "secret" or "protected" gardens. Fire, of course, was the element most associated with alchemical processes and a widely used

45. Pantheus's work contains several designs for furnaces, as does George Baker's translation of Conrad Gesner's work on distillation, *The newe Iewell of Health, wherein is contayned the most excellent Secretes of Phisicke and Philosophie, devided into fower Bookes* (London, 1576).

symbol of transformation and refinement. Finally, flowers appear throughout alchemical treatises and emblem books. The most common, the rose and the lily, were used to indicate color changes of the alchemical process and to differentiate between masculine and feminine principles.

In the four alchemical conversations that took place in the second half of 1584 and the beginning of 1585, these symbols were accompanied by additional symbols of alchemical processes and transformations. On 30 April 1584, for example, seven men in white priestlike robes (a personification of the seven planets) traveled up a hill carrying a tankard of transparent bone.[46] A partridge appeared (birds signified the union of masculine and feminine principles), but it had legs of different lengths. One of the men took the partridge and plucked most of its feathers off before cutting the longer leg. The partridge retained its wing feathers, however, and flew away (flight was a symbol of material volatility). Next the men arrived in a garden planted with roses and fruit trees (symbols of fertility and change). Inside there was a large white bird that flew about and beat down the bushes and trees with his wings. The bird attacked one of the men, who retaliated by cutting the bird's carcass into two parts and throwing them into the river.

Despite these setbacks, the men continued on the bridge and came to a round building with no windows that was nonetheless "bright." "It is a frame," Kelly explained, "made as though the 7 planets moved in it." Within the room was a suit of white armor, which one of the men donned before kicking the moon and turning it into a powder (powders were often used in alchemical processes). The men traveled on and came to a crowned priest (the masculine principle) who gave them a staff and a bottle before sending them on their way. They continued on to a great wilderness full of naked wild men. Finally, they reached a great river on the very top of the hill with "a rich tower all of precious Stones." Here the men filled their tankard, and one said, "This was my coming, and should be my return." Kelly reported that all seven men reappeared at the bottom of the hill, where they dug into the earth and removed a lion's carcass, which they revivified by giving it a drink from their tankard.

The final alchemical conversation shared by Dee and Kelly took place on 14 January 1585.[47] This conversation is more symbolically elusive than the others, and also more confusing. Despite these drawbacks, the conversation was probably of utmost importance to Dee, because it

46. *TFR*, pp. 111–114.   47. Ibid., pp. 355–361.

seemed to present the story of creation in alchemical terms. The angel Levanael appeared wearing an ash-colored veil (possibly a reference to the alchemical fire) in a garden full of fruit (representing the closed alchemical vessel). On a mound in the center of the garden stood a house that was perfectly round outside, but square inside. Four round windows and four doors faced in the cardinal directions. The north door was black, the east door was white, the south door was red, and the west door was green – the four dominant colors used in alchemical treatises. Within the east door, a fire burned. Within the west door, streams and rivers flowed. The south door was full of smoke, and the north door contained grey dust. Dee labeled these doors (respectively): fire, water, air, and earth. Lavanael left the house and met seven men (symbolic of the seven planets) vainly digging in the side of the hill. Levanael broke into the images at this point and told Dee and Kelly that "the Earth is a Monster with many faces: and the receptable of all variety. Go home, stand not idle. Provide by Arts for the hardnesse of Nature. . . ."

After a break, Dee and Kelly resumed their conversation with Levanael. The seven men were still digging, but in the process they were destroying their tools, complaining all the while that their purpose was unknown. A smith appeared who fixed their tools and made their work easier. One workman found a white substance, described as being like alabaster or salt. The workmen continued to dig until the earth was removed all the way to the round house's foundation. The workmen halted, unsure of how to proceed, lest they destroy it. Lavanael instructed them to continue but asked them to use their hands to dig. When the men uncovered the house's foundation, it was as clear as glass and burned with an interior fire. "[K]ernels of apples" burst from the house's north door, along with slime, "water thinner than slime," and pure water. "Stuff like yellow earth, which the fire wrought out of the black earth" appeared, and pure water seeped into the yellow earth. The fire receded, and creatures came forth. Levanael informed Dee and Kelly: "Here is Creation, and it is the first."

Levanael took a lump of earth and broke it into six round balls. He mixed yellow earth and water in an iron vessel, and the substance took on a greenish cast. Levanael, satisfied with his work, said, "Thou art strong, and wilt beget a strong Child" – a clear reference to the combination of masculine and feminine principles to yield an androgynous principle. When Levanael removed the substance from the iron vessel, it was transformed into gold. Lavanael repeated the process with another ball of earth, saying, "Corruption is a thief, for he hath robbed thee of thy best Ornaments, for thou art weaker in the second." When the second parcel of earth was removed, it was silver. The remaining balls of earth yielded copper, iron, and tin – the remaining metals in the alchemical hierarchy.

The angel's alchemical lessons continued, but they diminished in detail. Despite the confused allegories and murky meanings, it is clear that alchemy played a prominent role in the conversations. Increased alchemical knowledge would have been a strong incentive for Dee to continue the conversations, as would the angels' assurances that he was being "perfected" through his communication with them so that he could, in turn, restore the flawed human knowledge of nature and then perfect nature itself.[48] This relationship between the reformation of natural philosophy and the restitution of nature was strengthened when the angels told Dee that the true cabala of nature would enable him to attain mastery over nature and purify the earth as the final days unfolded.[49] In the *Monas hieroglyphica*, Dee had already come to believe that his synthesis of the human arts and sciences, especially those like alchemy and cabala which linked the perfect celestial world with the imperfect terrestrial world, would enable him as a natural philosopher to "effect a healing of the soul and a deliverance from all distress" and also empower him to gain "great wisdom, power over other creatures, and large dominion."[50]

48. Dee, *AWS* II:372.   49.   *TFR*, p. 65.
50.   Dee, *MH*, pp. 199 and 217.

Epilogue

Unfortunately, the religion that promised to reconcile all the people of the earth, the alchemical art that would have resolved debates between ancient and modern practitioners, and the *medicina dei* based on the reformed sciences of cabala and alchemy were not enough to heal the rift between John Dee and Edward Kelly. A few scattered angel conversations survive from the period after the angels revealed their reformed alchemy in 1585, and they indicate Kelly's growing absorption with his independent laboratory work and his efforts to fulfill the alchemical requests of noble patrons in Trebona and Prague. In 1589, Dee and Kelly parted company. The events of the interim period have been largely lost to us, so our questions about the alchemy described so briefly in 1585 remain unanswered, as do those about Dee's state of mind in 1588, when the marvelous year mentioned in the prophecies failed to usher in a complete restitution of the cosmos.

Dee's ongoing commitment to the angels after the dissolution of his partnership with Kelly is the most striking testament to his faith in their wisdom. Almost two decades later, he was still talking to the angel Raphael with the assistance of his last known scryer, Bartholomew Hickman (Figure 2). Perhaps Dee decided that his initial interpretation of the angels' prophecies about the year 1588 were faulty, and that the restitution of nature would actually occur in 1688 rather than 1588. During his last extant conversation, Raphael comforted Dee about his philosophical studies, confirmed the merits of the angelic revelations, and promised him a long life. Dee's finances, as ever, were, a problem, and there are signs in the angel conversations that Dee was peddling his angelic revelations in exchange for money. Dee asked Raphael about buried treasure and the actions of a thief named Thomas Webster on behalf of a client who had come to the natural philosopher for assistance.[1] Dee even considered entering the service of the earl of Salisbury and was willing to let James I's council know of his "beggary."[2] After

---

1. *TFR*, pp. *36 and *39.   2. Ibid. p. *32.

Dee's death, astrologer Richard Napier recalled that he was forced to sell some of his precious books to make ends meet.[3]

The image of Dee in his final years is far different from that of the confident young man enjoying the intellectual riches of Louvain or Paris, or even the vocal prophet-philosopher of Prague. The older Dee was beset with financial problems and plagued with doubts and worries. He had fallen on hard times, and his former patrons in England – such as the earls of Leicester and Pembroke, and William Cecil – were dead. No help arrived from Prague or Trebona, and ultimately Dee's repeated requests for increased livings and responsibilities came to nothing. On a more personal level, he was concerned for the welfare of his few surviving children, especially his son Arthur, who returned from his medical studies in Basel only to depart for Russia to become a physician to the tsar after his father's death.[4]

Dee was pacified in this troubled time by the angels' assurances that his work with them would be remembered for years to come. The angels once told Dee: "Whilest [sic] heven endureth and earth lasteth, *never shall be razed out the Memorie of these Actions*."[5] In this, as in the prophecies concerning the queen of Scots and the Spanish Armada, the angels' remarks proved true. Raphael told Dee that, when he died, people would marvel "that such an one was upon the earth, that God by him had wrought great and wonderful Miracles in his service." In addition, Raphael promised Dee, who had struggled throughout his life to improve the human condition through a more perfect understanding of the Book of Nature, that "thou shalt die with fame and memory to the end."[6]

Despite the infamy of Dee's angel conversations, however, they have remained an enigma. They can only become less enigmatic when placed firmly within the context of his life, work, and times. Though an advocate of the mathematical sciences that were to become increasingly im-

3. Roberts and Watson, p. 57.
4. *TFR*, p. *36; Roberts and Watson, pp. 63–64. More information on Arthur Dee can be obtained in the work of John H. Appleby: "Arthur Dee and Johannes Bánfi Hunyades: Further Information on Their Alchemical and Professional Activities," *Ambix* 24 (1977): 96–109; "Some of Arthur Dee's Associations before Visiting Russia Clarified," *Ambix* 26 (1979): 1–15; "Dr. Arthur Dee: Merchant and Litigant," *Slavonic and East European Review* 57 (1979): 32–55. See also N. A. Figurovski, "The Alchemist and Physician Arthur Dee (Artemii Ivanovich Dii): An Episode in the History of Chemistry and Medicine in Russia," *Ambix* 13 (1965): 35–51.
5. Dee, *AWS* II:247; Dee's emphasis.   6. *TFR*, p. *35.

portant in the centuries after his death, Dee was nonetheless immersed in the intellectual traditions passed down to him from the Middle Ages as well as those surrounding him in sixteenth-century Europe. These included the eschatological views of Guillaume Postel, the occult philosophies of Agrippa and Trithemius, the chemical medicine of Paracelsus, and the Christian cabala of Reuchlin. All helped Dee to formulate his own Christian natural philosophy, in which the word and works of God could be mastered, partially through study and contemplation, and completely through the gift of divine revelation and angelic communication.

A careful and comprehensive examination of Dee's angel conversations reveals that many of our preconceptions about them have been misleading. Neither secret, nor pagan, nor overtly magical, Dee's conversations resonate with a wide range of early modern contexts. R. J. W. Evans perceptively alluded to the context of Dee's angel conversations in his study of the court of Rudolf II. After analyzing the secular and clerical interest in Dee and his angels at the very highest levels of the Catholic and Rudolphine hierarchies, Evans concluded that Dee was an important and dangerous figure in the eyes of the papacy because the angels' revelations were "both *relevant* and *meaningful* to his [Dee's] Prague audience."[7] Anxieties expressed by Dee's contemporaries did not focus on whether the angel conversations were true or false. Rather, witnesses were worried about how Dee and his angelic prophecies could be managed, contained – perhaps even coopted – by existing religious and political hierarchies.

We have come to see John Dee through a darker glass than that provided by his own time. The dynastic change from Tudor to Stuart, the deaths of so many former patrons, and Dee's continued adherence to the idealistic aspirations of post-Reformation eschatologists all contributed to his declining prestige. Once praised as one of the most learned men in his native land and received by foreign emperors, kings, and universities as a distinguished member of the European intellectual community, Dee soon slid into ridicule and obscurity.

Dee's mysterious last days and the confusing events following his death have diminished the angel conversations and their role in his natural philosophy. Even the date of his death is now unclear. While scholars have traditionally dated it as December 1608, Julian Roberts and Andrew Watson have supplied persuasive marginalia to support a later date of 26 March 1609.[8] The absence of factual information is

7. R. J. W. Evans, *Rudolf II and His World* (Oxford: Clarendon Press, 1973), p. 223.
8. Roberts and Watson, p. 60.

emblematic of a much greater loss, however, for we lack a sense of the years preceding Dee's death as well. The few transcripts of angel conversations now extant are the greatest, yet still inadequate, access into Dee's life during his struggles with poverty, old age, and disappointment when the angels' promises of earthly rewards failed to materialize. During these years Dee published only apologies and supplications. Other work he might have completed, library purchases he might have made, and correspondence with European natural philosophers, have all been lost, leaving historians with little information upon which to draw. References in Dee's library books do, however, point to the identities of his final intellectual associates: alchemists John Pontois and Patrick Saunders.[9] Both men won Dee's confidence, and Pontois appears to have been especially favored.

Saunders was probably an alchemical assistant, for his name appears in connection with a loan of some of Dee's alchemical books to Henry Percy, the "Wizard Earl" of Northumberland.[10] Saunders later practiced medicine in London, and drew the unfavorable attention of the Royal College of Physicians in 1613. He was prohibited from practicing what Roberts and Watson describe as "Polish and German methods," which may indicate that Saunders learned Paracelsian ideas from Dee and through his library collection. Saunders was allowed to resume his medical practice after receiving a degree from the University of Franeker in 1619. Elias Ashmole knew Saunders and his activities, describing him as an astrologer and physician as well as a scryer with a reputation for seeing visions in crystals.[11] It is not surprising, in light of this description, that Saunders found a place with Dee.

Pontois was first mentioned in Dee's private diary in 1592.[12] The angels approved of him, and in 1607 Dee was preparing to embark on another journey – at the venerable age of sixty-five – accompanied by Pontois. The angel Raphael described Pontois as "a very honest and well-disposed young man," who would prove to be Dee's "greatest comfort and special ayd, next unto the Almighty."[13] Dee was concerned for the safety of his library during the journey, and Raphael told him to let Pontois take care of it. John Selden, in *De dis Syris* (1629), suggested that Dee bequeathed the control of the library to Pontois rather than to

9.   See ibid. pp. 57–68 for their account of the dispersal of Dee's books and for more information concerning his relationships with the people mentioned in this paragraph.
10.   Ibid. pp. 58–59.   11.   Ibid. p. 62.
12.   Dee, *PD*, p. 41; he refers to Pontois as "Ponsoys."
13.   *TFR*, pp. *37–*38.

his own son, Arthur Dee.[14] In a letter to Elias Ashmole dated 26 December 1672, Anthony Wood reported that Dee had died in a house in Bishopsgate Street in London, the same street (perhaps even the same house) where John Pontois lived.[15] In 1619 Pontois traveled to Virginia, where he became a prominent figure in colonial politics and administration. It is not known whether he took any of Dee's books with him, but Pontois returned to England in 1624 and died shortly thereafter.[16]

After Pontois's death in 1624, the remains of Dee's great library were formally sold.[17] But what of the angel diaries, the holy table, and other objects used in the angel conversations? Who possessed them after Dee's death? Were they, too, sold in Pontois's estate sale? Some items – most notably the ring and lamine – were no doubt buried with Dee. According to statements made by Patrick Saunders and a London grocer, Thomas Hawes, some other items were in the possession of John Pontois prior to his death.[18] In depositions taken during the settlement of Pontois's estate, Saunders described Pontois's double study, which contained some of Dee's books, a few old trunks, "a Table . . . which Pountys [sic] rec[k]oned to be of great value," and a "stone or Jewell of Cristall." Hawes's description of the double study was more detailed. The grocer referred to "a certain round flat stone like a Christall which Pountis said was a stone which an Angell brought to doctor dye wherein he did worke and know many strange things." The valuable table was also more fulsomely described as a "Table with Caracters and devises uppon it which was kept covered by the said Pountis in the innermost of the upper studdyes."

As for the angel diaries, the volumes later printed by Casaubon were purchased from Pontois's estate by Sir Robert Cotton along with the holy table.[19] The fate of the remainder, those unpublished until Christopher Whitby's edition, has remained more conjectural. Another man, John Woodall, has been linked to Pontois's death. Elias Ashmole recorded that it was through Woodall that the diaries came into the hands of the London confectioner Mr. Jones.[20] Woodall was a Paracelsian

14. John Selden, *De dis Syris* (Leiden, 1629), p. 88; this remark is discussed in Roberts and Watson, p. 60.
15. Roberts and Watson, p. 60.   16. Ibid., p. 61.
17. Ibid. pp. 64–65.   18. Ibid., pp. 61–62.
19. Casaubon, "Preface," *TFR*, sig. [F1ᵛ].
20. For Ashmole's account of the manuscripts' history, see Dee, *AWS* II:2–4. For information on Woodall, see Roberts and Watson, pp. 61 and 63; Allen G. Debus, "John Woodall, Paracelsian Surgeon," *Ambix* 10 (1962): 108–118; Allen G. Debus, *The English Paracelsians* (1965) (New York: Franklin Watts, 1966), pp. 99–101.

physician and a member of the Virginia Company. When Pontois departed for the colonies, he left the keys to his studies, counting house, and stillhouse with Woodall.

Between 1624 and 1672 we know nothing of Dee's diaries. Elias Ashmole came to possess them from the widow of confectioner Jones, who had purchased a wooden chest from a "parcell of the Goods of Mr John Woodall Chirurgeon." The chest contained Dee's earliest angel diaries in a secret compartment.[21] Though Ashmole credits Woodall with purchasing the item "after Dr Dee's death, when his goods wer exposed to Sale," it is also quite possible that Woodall made off with the diaries – either wittingly or unwittingly – after Pontois's death. Once in the Jones's household many folios from the angel diaries were destroyed. Finally, the diaries passed from Ashmole's hands into the collections of the British Library through Sir Hans Sloane, although the exact delineation of this transfer is not clear.[22]

Dee's death and the dispersal of his possessions failed, as the angels had promised, to delete his angel conversations from memory. Once selections from the angel diaries were published by Meric Casaubon in 1659, Dee's activities became public knowledge. Tantalizing glimpses of a sustained interest in scrying and angelology after 1659 suggest that the seventeenth century was no less interested in the occult features of the natural and supernatural worlds than Dee had been. Not everyone remembered Dee's efforts for the same reasons, or was interested in the same aspects of the angel conversations, however. Some scholars, like Casaubon, were committed to suppressing and discrediting Dee's angel conversations so that others would be discouraged from scrying. Natural philosophers, on the other hand, were interested in the methods and theories that provided the theoretical basis for the conversations, and some might have been intrigued by Dee's references to the prophet Elias's return. Generally, people were intrigued by the angels, as well as by the ritualistic aspects of the conversations.

The subsequent interest in Dee's angel conversations deserves further consideration. Although a detailed examination of seventeenth-century attitudes promises to reveal new information about experimental practices, the status of the occult sciences, and the growing belief in the power of reason over the power of revelation, such an undertaking is beyond the limits of this study. One of the most prominent seventeenth-century natural philosophers to express an interest in scrying, for exam-

21. Ashmole's introduction in Dee, *AWS* II:2–4.
22. Christopher Whitby discusses the murky chain of events that might have led from Ashmole to Sloane. See Whitby, *AWS* I:3–7.

ple, was the chemist Robert Boyle (1626–1691). Boyle, by his own admission, was "tempted" to try his hand at scrying.[23] In autobiographical remarks dictated to his friend Gilbert Burnet, Boyle reported several scrying episodes, including one involving a cleric who offered to show Boyle "strange representations" in a glass of water.[24] Though Boyle did not accept the offer, he later felt "the greatest Curiousity . . . in his life" when a gentleman brought him "an Ordinary double convexe glasse" in which to see spirits.[25] Boyle overcame his intense desire to scry and considered this denial of a potentially sinful activity his greatest moral achievement. Such a reaction is intriguing, especially in light of Boyle's belief, shared by Dee and Ashmole, that communication between a natural philosopher and the supernatural or spiritual parts of the cosmos would facilitate the genesis of the philosophers' stone.[26]

As observed, in Dee's work the philosophers' stone was linked to the restitution of nature as well as the figure of Elias. The role that Elias would play in the restitution of nature continued to exert a powerful influence on seventeenth-century natural philosophers, particularly those influenced by Paracelsus. In central Europe, Gerhard Dorn, Johannes Montanus and Alexander von Suchten publicized Paracelsus's belief that "Elias artista" would restore the arts and sciences before the final days.[27]

---

23. For a discussion of this incident see Michael Hunter, "Alchemy, Magic, and Moralism in the Thought of Robert Boyle," *British Journal for the History of Science* 23 (1990): 387–410. I am indebted to Margaret G. Cook of the University of Calgary for bringing this incident to my attention.
24. Ibid., p. 390.   25. See ibid., p. 391.
26. See ibid., pp. 391 and 396–397.
27. Herbert Breger, "*Elias Artista*—a Precursor of the Messiah in Natural Science," in *Nineteen Eighty-Four: Science between Utopia and Dystopic*, ed. Everett Mendelsohn and Helga Nowothy (New York: D. Reidel Publishing Company, 1984), p. 56. Dee owned a particularly rich collection of works by both Dorn and von Suchten. In the case of Dorn, Dee's books included: two copies of the *Lapis metaphysicus* (1570), Roberts and Watson, #1447-Fr and #2285; *Monarchia physica* (Basel, 1577), Roberts and Watson, #1512-T; *Chimistici artificii partes tres* (1568), Roberts and Watson, #1524-T; *Clavis chimistica philosophiae* (Lyon, 1567), Roberts and Watson, #1525-T. The works of von Suchten in Dee's library consisted of three copies of his *Libellus de secretis antimonii* (Augsburg, 1570), Roberts and Watson, #1454-T, 1529-T, and 2276-T. In this work, von Suchten makes specific references to Elias, as Breger notes. Dee criticized Dorn for using his *monas* symbol without acknowledgment ("sine grata nostri mentione") on the title page of his copy of part one of the *Chimistici artificii*, now New York Society Library no. 86.

Though Herbert Breger notes that the "Elias artista school of thought became widespread in the last three decades of the sixteenth century," it is evident that it did not cease to be important after Dee's death.[28] Nor was interest in Elias confined to followers of Paracelsus. Both Jan Baptista van Helmont (ca. 1577–1644) and his son, Francis Mercury van Helmont (1618–1699), spoke of Elias in their works. Ultimately, Breger credits the gradual disappearance of Elias from natural philosophical texts to the "decline in the belief that the world was soon to end . . . and the rise of a new mechanical and rational philosophy."[29] Scholars have not, as yet, precisely studied Elias's disappearance, but this promises to be a fruitful avenue of inquiry.

Not all natural philosophers – not even all occult philosophers – were proponents of the blend of eschatology and natural philosophy found in Dee and Paracelsus, however, and some were deeply skeptical of Dee's angel conversations. Meric Casaubon, now known almost exclusively for his role in the publication of the angel conversations, found much to fault in Dee's reliance on spiritual advisers and direct revelation. Casaubon's preface suggests that, though the tenets of Christian natural philosophy were still known, people were unsure of its methods and results. In Casaubon's estimation, for example, Dee was undoubtedly "a very free and sincere Christian," but Casaubon was not convinced that he really deserved the title of mathematician, despite his piety and devout prayers.[30] Ultimately, Casaubon's low opinion of Dee's angel conversations stemmed from his belief that he had been deceived by bad spirits masquerading as good angels. Casaubon hoped that the publication of the sordid details of Dee's angel conversations would convert atheists to a belief in the spiritual world, moderate the opinions of advocates of prophecy and revelation (such as the Anabaptists) who relied too much on spirits for answers to their dilemmas, and encourage people to appreciate the power of prayer. In addition, the publication would reorient people who consulted astrologers and conjurers – as well as those who read about such dangerous ideas in books.[31]

Despite Casaubon's fervent desire that people would abstain from attempting to contact "false lying Spirits," *A True and Faithful Relation* seems only to have encouraged such activities.[32] A set of manuscript diaries now in the British Library records the activities of a group that gathered between 24 July 1671 and 18 December 1688 to converse with angels through a crystal. The dates of the last conversations suggest that

28. Breger, "*Elias artista*," p. 57.   29. Ibid., p. 62.
30. Casaubon, "Preface," TFR, sigs. Aʳ – Iᵛ, sig. Dʳ.
31. Ibid., sigs. [H1ʳ⁻ᵛ].   32. Ibid., sig. Dᵛ.

the group initially became involved with the angels after reading Casaubon and discovering that there was some ambiguity in Dee's mind as to whether the much prophesied year of cosmic restitution was actually 1588 or 1688. Like Dee, the group employed a scryer "E. R[orbon]," and two other regular participants are mentioned: "R. O." who occupied Dee's role in the conversations, and another referred to at times as "E. C." and occasionally as "Brother Collings."[33] The angels the group conversed with were identical to Dee's, and the angels Madimi and Galuah frequently appeared – two of Dee's most idiosyncratic angels. Although these later conversations superficially resemble Dee's, however, the practitioners were not deeply engaged with natural philosophy but were primarily concerned with their financial security, and there was a great deal of interest in the recovery of buried riches.

Although a wider circle of readers was exposed to his attitudes toward the contents of the angel diaries through Casaubon's publication of *A True and Faithful Relation*, his criticisms of the work were not nearly as dismissive and damning as those of the natural philosopher Robert Hooke. It is evident from Hooke's reference to the "many Discourses" he overheard among his friends at the Royal Society that the angel conversations were a popular topic there.[34] In a lecture given to the Royal Society on the subject, Hooke referred scathingly to the angel conversations as "Dr. Dee's Delusion," though he considered Dee to have been "an extraordinary Man, both for Learning, Ingenuity and Industry."[35] Hooke rebuked Casaubon as well, stating that *A True and Faithful Relation* was "extravagant," despite the fact that he agreed that the documents were a valuable weapon against atheists, "Enthusiasts, who altogether depend upon new Revelations," and those unfortunate enough to seek out witches, conjurers, and astrologers.[36] Nevertheless, Hooke, like Dee and Casaubon before him, found something of interest in the diaries. He believed that the conversations should not be accepted at face value, but were a form of cryptography – one of Hooke's own passions – which hid valuable political information that Dee had gathered for Queen Elizabeth.[37]

The angel conversations took on a different cast outside the world of scholars and natural philosophers, and hints of Dee's interest in angels appeared in several seventeenth-century dramas. Dee's troubled and star-crossed life may have lent itself to dramatization even before the publication of the angel conversations. Frances Yates has suggested that sto-

33. See British Library, Sloane MSS 3624–3628, passim.
34. Robert Hooke, *The Posthumous Works* (London, 1705), p. 205.
35. Ibid., pp. 203–204.   36. Ibid., p. 204.   37. Ibid., pp. 206–207.

ries of Dee's alchemical and angelic activities provided the model for Ben
Jonson's play *The Alchemist* (1610).[38] This seems likely, given the plot,
yet Casaubon had not printed the diaries at that early date. Jonson
would have had to draw on popular lore for his information, and there
are no precise references to Dee's angel conversations in *The Alchemist*.
After Casaubon's edition of excerpts from the angel diaries, however,
Samuel Butler was able to include more detailed information about Dee's
activities in his satirical poem *Hudibras* (1664). There Butler describes a
Rosicrucian cunning-man and natural philosopher, Sidrophel:

> H'had read Dee's prefaces before,
> The Dev'l and Euclid o're and o're
> And, all the intregues 'twixt him and Kelly,
> Lescus [Laski] and th'Emperor [Rudolph] would tell yee. . . . [39]

Later in the poem, Butler found a way of expressing his own opinion of
the angel conversations:

> Kelly did all his feats upon
> The devil's looking-glass, a stone,
> Where playing with him at Bo-peep
> He solv'd all problems ne'r so deep.[40]

Butler was well versed in the contents of *A True and Faithful Relation*,
though he dismisses the conversations as a form of charades produced
by Kelly.

The protracted metamorphosis of Dee's angel conversations described
here – a transformation which led from his own conception of them as
the pinnacle of Christian natural philosophy and the way to practice
natural philosophy at the end of the world, to the moment when Casaubon published them to discourage conjurers, to Hooke's theory that they
were a form of cryptography – has the potential to tell us something
about other changes that affected natural philosophy. "Dr. Dee's delusion" was not how the angel conversations were described in his own
time; this description surfaced in the century after his death, when the
effects of humanism and the Reformation were being felt and expressed
in very different ways. In the seventeenth century, Dee's emphasis on the
Book of Scripture and the power of divine revelation was replaced by an
emphasis on the Book of Nature and the power of human reason.
Because Dee's goals were shrouded in eschatological prophecies, alchem-

38. Frances A. Yates, *The Occult Philosophy in the Elizabethan Age* (London:
     Routledge and Kegan Paul, 1979), pp. 161–162.
39. Samuel Butler, *Hudibras*, ed. John Wilders (Oxford: Clarendon Press,
     1967), II.iii.235–238.
40. Ibid., pp. 631–634.

ical parables, and cabalistic allusions, they were identified as part and parcel of natural philosophy's bookish and allegorical past rather than its experimental and observational future.

More recently, Dee's angel conversations have been cast out of natural philosophy entirely, and little space has been reserved for Christian natural philosophers like John Dee. Yet Christian natural philosophy was a vital part of the early modern period's complex intellectual life. The men who advocated and practiced it – such as Agrippa, Paracelsus, and Postel – were among the most lauded scholars of their time. Keith Thomas, in his examination of early modern prophets and their activities, warned that it is "not enough to describe such men as lunatics." Instead, Thomas encouraged historians "to explain why their lunacy took this particular form."[41] In the case of Dee, it was his challenge to practice natural philosophy with a corrupted text – the Book of Nature – which encouraged him to find a new, incontrovertible source of natural philosophical information. John Dee's angel conversations came into being at a particular moment and time, and blended into a unique, but comprehensible, attempt to practice natural philosophy at the end of the world.

41. Keith Thomas, *Religion and the Decline of Magic* (New York: Scribner's, 1971), p. 149.

Select Bibliography

Works Written or Annotated by John Dee

Ambrosius, Theseus. *Introductio in Chaldaicam linguam, Syriacum atque Armenicam et decem alias linguas.* 1539. Roberts and Watson, #1620. Oxford, Bodleian Library 4° A 55 Art. Seld.

Avicenna. *Compendium de anima....* Venice, 1546. Roberts and Watson, #395. Oxford, Bodleian Library 4° A 40 Art. Seld.

Azalus, Pompilius. *De omnibus rebus naturalibus.* Venice, 1544. Roberts and Watson, #134/B188. Cambridge, Emmanuel College 304.I.54.

Dee, John. "Compendious Rehearsal." In *Autobiographical Tracts of Dr. John Dee, Warden of the College of Manchester*, pp. 1–45. Ed. James Crossley. London: Chetham Society, 1851.

————. *Diary, for the Years 1595–1601, of Dr. John Dee, Warden of Manchester from 1595 to 1608.* Ed. John Eglinton Bailey, F. S. A. Privately printed, 1880.

————. *The Enochian Evocation of Dr. John Dee.* Ed. and trans. Geoffrey James. Gillette, NJ: Heptangle Books, 1988.

————. *The Heptarchia Mystica.* Annotated by Robert Turner. Wellingborough, Northants.: Aquarian, 1986.

————. "Introductory Epistle." *Ephemeris Anni 1557. Currentis Iuxta Copernici et Reihnali Canones.* Comp. John Feild. London, 1556.

————. *John Dee's Actions with Spirits.* Dissertation, University of Birmingham, 1981. Ed. and foreword by Christopher Whitby. 2 vols. New York: Garland, 1988. (*See also* Whitby)

————. *A Letter, Containing a Most Briefe Discourse Apologeticall, with a Plaine Demonstration, and Fervent Protestation, for the Lawfull, Sincere, Very Faithfull and Christian Course, of the Philosophicall Studies and Exercises, of a Certaine Studious Gentleman: An Ancient Servaunt to Her Most Excellent Maiesty Royall (1599).* The English Experience, no. 502. Amsterdam: Theatrum Orbis Terrarum, 1973.

————. "Letter to Sir William Cecil, 16 February 1562/63." Ed. R. W. Grey. *Philobiblon Society Bibliographical and Historical Miscellanies* 1 (1854): 1–16.

————. *The Mathematicall Praeface to the Elements of Geometry of Euclid of Megara (1570).* Ed. Allen G. Debus. New York: 1975.

———. "A Playne Discourse and Humble Advise . . . concerning ye needful reformation of ye vulgar kallendar." Bodleian Library MS Ashmole 1789, ff. 1–40.

———. *Private Diary of Dr. John Dee.* Ed. J. O. Halliwell. Camden Society, vol. 19. London, 1842.

———. "Propaedumata aphoristica." Ed. and trans. Wayne Shumaker. *John Dee on Astronomy.* Foreword by J. L. Heilbron. Berkeley: University of California Press, 1978.

———. "A Supplication to Queen Mary, 1556." *John Dee's Library Catalogue,* pp. 194–195. London: The Bibliographical Society, 1990.

———. "A Translation of John Dee's 'Monas hieroglyphica' (Antwerp, 1564), with an Introduction and Annotations." Ed. and trans. C. H. Josten. *Ambix* 12 (1964): 84–111.

———. *A True and Faithful Relation.* Ed. and foreword by Meric Casaubon. London, 1659.

Dionysius. *De Mystica Theologica.* Rome, 1525. Roberts and Watson, #271. Cambridge, Cambridge University Library H*8.22(c).

———. *Opera.* Commentary by Jacques Lefèvre d'Etaples. Venice, 1556. Roberts and Watson, #975. Cambridge, Magdalene College D.7.10.

Dorn, Gerhard. *Chemistici artificii partes tres* (1568). Roberts and Watson, #1524-T. Part I, New York Society Library no. 86; Parts II–III, possibly Yale University, Beinecke Library Yna31 569d.

Elias Levita. *Composita Verborum et Nominum Hebraicorum.* Trans. Sebastien Muenster. Basel, 1525. Roberts and Watson, #1594-Fr. Cambridge, Cambridge University Library K*6.36(F).

Gesner, Conrad. *Bibliotheca universalis. Epitome* by Josias Simler, 1574. Roberts and Watson, #282. Bodleian Library, Oxford, Arch. H.c.7.

Happellius, Wigandus. *Linguae Sanctae Canones Grammatici.* Basel, 1561. Roberts and Watson, #1600-Fr. Cambridge, Cambridge University Library K*6.37(F).

Henricus de Herph. *Theologiae Mysticae.* Cologne, 1556. Roberts and Watson, #223. Cambridge, Cambridge University Library H*1.10(B).

Hermes Trismegistus. "Asclepius." *Index eorum, quae hoc in libro habentur. Iamblichus, de mysteriis Aegyptiorum . . . ,* pp. 125–[141ʳ]. Ed., trans., and comp. Marsilio Ficino. Venice: Aldus, 1516. Roberts and Watson, #256. Folger Shakespeare Library BF 1501 J2 Copy 2 Cage.

Iamblichus. "De mysteriis Aegyptiorum, Chaldaeorum, Assyriorum." *Index eorum, quae hoc in libro habentur. Iamblichus, de mysteriis Aegyptiorum . . . ,* pp. 1–23. Ed., trans., and comp. Marsilio Ficino. Venice: Aldus, 1516. Roberts and Watson, #256. Folger Shakespeare Library BF 1501 J2 Copy 2 Cage.

Josten, C. H., ed. and trans. "An Unknown Chapter in the Life of John Dee." *Journal of the Warburg and Courtauld Institutes* 28 (1965): 223–257.

Maginus, Johannes Antonius. *Ephemerides Coelestium Motuum Jo. Antonii Magini Patavini, Ad Annos XL. Ab Anno Domni 1581. Usque Ad Annum 1620. Secundum Copernici Hypotheses, Prutenicosq[ue]; Canones, Atq[ue];*

Iuxta Gregorianam Anni Correctionem Accuratissime[m] Supputatae. Ad Longitudinem Gr. 32.30'. Sub Qua Inclyta Urbs Venetriarum Sita Est. Addita Est Eiusdem in Stadium Animadversio, Qua Errores Eius Quamplurimi Perpenduntur. Item Tractatus Quatuor Absolutissimi, Nempe Isagoge in Iudiciariam Astrologiam, De Usu Ephemeridum, De Annuis Revolutionibus, & De Stellis Fixis. Venice, 1582. Bodleian Library Ashmole MS 488.

Manilius, Marcus. *Astronomica.* Basel, 1533. Roberts and Watson, #251. University College, London Ogden A.9.

Muenster, Sebastien. *Messias Christianorum et Iudaeorum Hebraicè & Latinè.* Basel, 1539. Roberts and Watson, #1616. Cambridge, Cambridge University Library K*6.60.

———. ed. and trans. *Evangelius Secundum Mathhaeum in Lingua Hebraica, Cum Versione Latine.* Basel, 1557. Roberts and Watson, #1588-T.

Pagninus, Santes. *Hebraicarum Institutionum Libri IIII, Sante Pagnino Lucenis Authore, Ex R. David Kimhi Priore Parte . . .* [Par.], [1549]. Oxford, Bodleian Library 4° P 47 Art. Seld.

Pantheus, Joannes Augustinus. *Voarchadumia Contra Alchimiam: Ars Distincta Ab Archimia, & Sophia: Cum Additionisubs: Proportionibus: Numeris: & Figuris Opportunis Ioannis Augustini Panthei Veneti Sacerdotis.* Venice, 1530. Robert and Watson, #D16. London, British Library C.120.b.4(2).

Postel, Guillaume. *Cosmographia.* Basel, 1561. Roberts and Watson, #386. London, Royal College of Physicians D 49/6, 46f.

———. *De orbis terrae concordia libri quatuor* [Basel, 1544]. Roberts and Watson, #D18. London, Westminster Abbey Library O.163.

———. *De Originibus, Seu, de Varia et Potissimum Orbi Latino Ad Hand Diem Incognita, Aut Inco[n]syderata Historia Quu[m]totius Orientis, Tum Maxime Tartarorum, Persarum, Turcarum, & Omnium Abrahami & Noachi Alumnorum Origines, & Mysteria Brachmanum Retegente: Quod Ad Gentium, Literarumq[ue] Quib.utuntur, Rationes Attinet.* Paris, 1538. Roberts and Watson, #1623. Oxford, All Souls' College SR.17.d.3.

———. *De Originibus.* Basel, 1553. Roberts and Watson, #868. London, Royal College of Physicians D144/14, 21b.

———. *Linguarum Duodecim Characteribus Diferentium Alphabetum, Introductio, Ac Legendi Modus Longe Facilimus. Linguarum Nomina Sequens Proxime Pagella Offeret.* Paris, 1523. Roberts and Watson, #1623. Oxford, All Souls' College SR.17.d.3.

Proclus. *Primum Euclidis Elementorum Librum Commentariorum Ad Universam Mathematicam Disciplinam Principium Eruditionis Tradentium Libri Iiii.* Padua, 1560. Roberts and Watson, #266. Bodleian Library Savile W17.

Psellus, Michael. "De Daemonibus." *Index eorum, quae hoc in libro habentur. Iamblichus, de mysteriis Aegyptiorum . . .* , pp. 50–54. Ed., trans., and comp. Marsilio Ficino. Venice: Aldus, 1516. Roberts and Watson, #256. Folger Shakespeare Library BF 1501 J2 Copy 2 Cage.

Roberts, Julian, and Andrew G. Watson, eds. *John Dee's Library Catalogue.* London: The Bibliographical Society, 1990.

Scaliger, Julius Caesar. *Exotericarum Exercitationum Liber Quintus Decimus, de Subtilitate, Ad Hieronymum Cardanum*. Paris, 1557. Roberts and Watson, #476. London, Royal College of Physicians D129/2, 18c.

Stadius, Joannes. *Ephemerides Novae, Auctae et Repurgatae Ioannis Stadii Leonnouthensis Mathematici. Secundae Antverpiae Longitudinem. Ab Anno 1554. Usque Ad Annum 1600*. Cologne, 1570. Oxford, Bodleian Library Ashmole MS 487.

Trithemius, Johann. *De Septem Secundeis*. Frankfurt, 1545. Roberts and Watson, #678-T. Cambridge, Cambridge University Library Dd*4.5.

———. *Liber Octo Quaestionum, Quas Illi Dissolvendas Proposuit Maximilianus Caesar*. Cologne, 1534. Roberts and Watson, #897. Cambridge, Cambridge University Library H*15.9(F).

Manuscripts

Cambridge, Cambridge University Library
Add. MS 3544
MS Li.1.12
London, British Library
Add. MS 36674
Landsdowne MS 121
Sloane MS 8
Sloane MS 2544
Sloane MS 3188
Sloane MS 3189
Sloane MS 3190
Sloane MS 3191
Sloane MSS 3624–3628
Sloane MS 3846
Sloane MS 3849
Sloane MS 3851
Sloane MS 3884
Oxford, Bodleian Library
Ashmole MS 1790
Ashmole MS 1788
Bodley MS 908
e. Mus. MS 238
Rawlinson MS D253

Primary Sources

Agnelli, Giovanni Baptista. *Apocalypsis spiritus secreti*. London, 1566.
Agrippa, Henry Cornelius. *De occulta philosophia libri tres*. Basel, 1533.
———. *Agrippa, His Fourth Book of Occult Philosophy; Of Geomancy; Magical Elements of Peter De Abano; Astronomical Geomancy; the Nature of*

Spirits; Arbatel of Magick. Trans. Robert Turner. London: JC for John Harrison, 1655.

———. *Of the vanitie and uncertaintie of the Artes and Sciences.* Trans. James Sanford. London, 1569.

———. *Three Books of Occult Philosophy.* Trans. J[ohn] F[rench]. London, 1651.

Ashmole, Elias, comp. and ed. *Theatrum Chemicum Britannicum.* London, 1652.

Bacon, Roger. *Frier Bacon His Discovery of the Miracles of Art, Nature, and Magick. Faithfully Translated Out of Dr. Dees Own Copy by T. M. and Never Before in English.* London: Simon Miller, 1659.

———. *Roger Bacon's Letter Concerning the Marvelous Power of Art and Nature and Concerning the Nullity of Magic.* Trans. Tenney L. Davis. Easton, PA: Chemical Publishing Company, 1923.

Barrett, Francis. *The Celestial Intelligencer.* London, 1801.

Biringuccio, Vannoccio. *The Pirotechnia of Vannoccio Biringuccio.* Ed. and trans. Cyril Stanley Smith and Martha Teach Gnudi. New York: Dover, 1990.

Bostocke, R[obert?]. *The Difference betwene the auncient Phisicke . . . and the latter Phisicke.* London, 1585.

Brahe, Tycho. *Tychonis Brahe Dani Opera omnia.* Ed. J. L. E. Dreyer. 15 vols. Copenhagen, 1913–1929.

———. *His Astronomicall Coniectur of the New and Much Admired * Which Appered in the Year 1572.* London, 1632.

Butler, Samuel. *Hudibras.* Ed. John Wilders. Oxford: Clarendon Press, 1967.

Cardano, Girolamo. *The Book of My Life (De Vita Propria Liber).* Trans. Jean Stoner. London: J. M. Dent and Sons, 1931.

Charles, R. H., ed. "The Book of Enoch." *Apocrypha and Pseudepigrapha of the Old Testament,* 2:163–281. Oxford: Oxford University Press, 1913.

Copenhaver, Brian P., trans. and ed. *Hermetica: The Greek "Corpus Hermetica" and the Latin "Asclepius."* Cambridge: Cambridge University Press, 1992.

Davies, John. *The Poems of Sir John Davies.* Ed. Robert Krueger. Oxford: Clarendon Press, 1967.

Della Porta, John Baptista. *Natural Magick.* London, 1658.

Dionysius the Areopagite. *The Complete Works.* The Classics of Western Spirituality. London: SPCK, 1987.

Euclid. *The Elements of Geometrie of the most auncient Philosopher* EUCLIDE *of Megara.* Trans. H[enry] Billingsley. Preface by John Dee. London: John Daye, 1570.

Forman, Simon. "Magical Papers." British Library Add. MS 36674, ff. 47–56ᵛ.

Foxe, John. *Acts and Monuments.* 8 vols. New York: AMS Press, 1965.

Gesner, Conrad. *The newe Iewell of Health. wherein is contayned the most excellent Secretes of Phisicke and Philosophie, devided into fower Bookes.* Trans. George Baker. London, 1576.

Giuntini, Francesco. *Speculum Astronomia.* Paris, 1573.

Hooke, Robert. *The Posthumous Works.* London, 1705.

Leowitz, Cyprian. *De conjunctionibus magnis insignioribus superiorum plane-tarum* (1564).

McLean, Adam, ed. *A Treatise on Angel Magic.* Magnum Opus Hermetic Sourceworks, no. 15. Grand Rapids, MI: Phanes Press, 1990.

Mayor, J. E. B., ed. *Early Statutes of the College of St. John the Evangelist in the University of Cambridge.* Cambridge: Cambridge University Press, 1859.

Meres, Frances. *Palladis Tamia.* London, 1598.

Muenster, Sebastien. *The Messias of the Christians and the Jewes.* Trans. Paul Isaiah. London: W[illia]m Hunt, 1655.

Pantheus, Giovanni. *Ars transmutationis metallicae.* Venice, 1518.

Paracelsus. *The Archidoxes of Magic of Paracelsus* (1656). Trans. R[obert] Turner. New York: Samuel Weiser, 1975.

———. *Samtliche Werke.* Ed. Karl Sudhoff. 14 vols. Munich: Oldenbourg, 1922–1933.

———. *Selected Writings.* Ed. Jolande Jacobi. Trans. Norbert Guterman. Bollingen Series, vol. 28. Princeton, NJ: Princeton University Press, 1988.

Pico della Mirandola, Giovanni. *Conclusiones Sive Theses DCCCC Romae Anno 1486 Publice Disputandae, Sed Non Admissae.* Introduced and annotated by Bohdan Kieszkowski. Geneva: Librairie Droz, 1973.

———. *Opera Omnia.* Basel, 1572.

Proclus. *A Commentary on the First Book of Euclid's Elements.* Trans., notes, and introduction by Glenn R. Morrow. Princeton, NJ: Princeton University Press, 1970.

Pucci, Francesco. *Lettere, documenti e testimonianze.* Ed. Luigi Firpo and Renato Piattoli. 2 vols. Opuscoli Filosofici Testi e Documenti inediti o rari. Florence: Leo S. Olschki, 1955–1959.

Purvis, J. S., ed. *The York Cycle of Mystery Plays.* London: SPCK, 1957.

Raleigh, Walter. *The History of the World.* London, 1614.

Reuchlin, Johannes. *De Verbo Mirifico.* Paris, 1552.

———. *On the Art of the Kabbalah, De Arte Cabalistica.* Trans. Martin Goodman and Sarah Goodman. Intro. by G. Lloyd Jones. New York: Abaris Books, 1983.

Selden, John. *De dis Syris.* Leiden, 1629.

Stowe, John. *Annals of England.* London, 1632.

Trithemius, Johann. *The Steganographia of Johannes Trithemius.* Ed. Adam McLean. Trans. Fiona Tait, Christopher Upton, and J. W. H. Walden. Edinburgh: Magnum Opus Hermetic Sourceworks, 1982.

Secondary Sources

Abbot, A. E. *Encyclopaedia of the Occult Sciences.* London: Emerson Press, 1960.

Abraham, Lyndy. *Dictionary of Alchemical Imagery*. Cambridge: Cambridge University Press, forthcoming.

Allen, Michael J. B. "The Absent Angel in Ficino's Philosophy." *Journal of the History of Ideas* 36 (1975): 219–240.

Ansani, Antonella. "Giovanni Pico Della Mirandola's Language of Magic." In *L'Hebreu au temps de la Renaissance*, ed. Ilana Zinguer, pp. 89–104. Brill's Series in Jewish Studies 4. New York: E. J. Brill, 1992.

Appleby, John H. "Arthur Dee and Johannes Bánfi Hunyades: Further Information on Their Alchemical and Professional Activities." *Ambix* 24 (1977): 96–109.

———. "Dr. Arthur Dee: Merchant and Litigant." *Slavonic and East European Review* 57 (1979): 32–55.

———. "Some of Arthur Dee's Associations before Visiting Russia Clarified." *Ambix* 26 (1979): 1–15.

Ashworth, William B. "Natural History and the Emblematic World View." In *Reappraisals of the Scientific Revolution*, pp. 303–322. Ed. David Lindberg and Robert S. Westman. Cambridge: Cambridge University Press, 1990.

Aston, Margaret E. "The Fiery Trigon Conjunction: An Elizabethan Astrological Prediction." *Isis* 61 (1970): 159–187.

Bailey, John E. "Dee and Trithemius's 'Steganography.'" *Notes and Queries* 5th ser., 11 (1879): 401–402 and 422–423.

Baker, Dora. *Giovanni Pico della Mirandola, 1463–1494: Sein Leben und sein Werk*. Dornach: Verlag am Goetheanum, 1983.

Ball, Bryan W. *A Great Expectation: Eschatalogical Thought in English Protestantism to 1660*. Studies in the History of Christian Thought, vol. 12. Leiden: E. J. Brill, 1975.

Barker, Peter. "Jean Peña and Stoic Physics in the Sixteenth Century." In *Recovering the Stoics*, ed. Ronald H. Epp. *Southern Journal of Philosophy* 23, supplement (1985): 93–108.

Barnes, Robin Bruce. *Prophecy and Gnosis: Apocalypticism in the Wake of the Lutheran Reformation*. Stanford, CA: Stanford University Press, 1988.

Barone, Robert William. "The Reputation of John Dee: A Critical Appraisal." Ph.D. diss., Ohio State University, 1989.

Bauckham, Richard, ed. *Tudor Apocalypse: Sixteenth Century Apocalypticism, Millenarianism, and the English Revolution*. Oxford: Sutton Courtenay Press, 1978.

Berefelt, Gunnar. *A Study on the Winged Angel. The Origin of a Motif*. Trans. Patrick Hort. Stockholm: Almquist and Wiksell, 1968.

Besterman, Theodore. *Crystalgazing: A Study in the History, Distribution, Theory, and Practice of Skrying*. London, 1924.

Bischoff, Guntriem G. "Dionysius the Pseudo-Areopagite, the Gnostic Myth." In *The Spirituality of Western Christendom*, ed. E. Rozanne Elder. Kalamazoo, MI: Cistercian Publications, 1976.

Blair, Ann. *The Theater of Nature: Jean Bodin and Renaissance Science*. Princeton, NJ: Princeton University Press, 1997.

Blau, J. L. *The Christian Interpretation of the Cabala in the Renaissance* (1944). New York: Columbia University Press, 1965.

Blumenberg, Hans. *Die Lesbarkeit der Welt.* 2 vols. Frankfurt am Main: Surhkamp, 1981.

Bono, James. "Reform and the Languages of Renaissance Theoretical Medicine: Harvey versus Fernel." *Journal of the History of Biology* 23 (1990): 341–387.

———. *The Word of God and the Languages of Man.* Madison: University of Wisconsin Press, 1995.

Borchardt, Frank L. *Doomsday Speculation as a Strategy of Persuasion: A Study of Apocalypticism as Rhetoric.* Studies in Comparative Religion, vol. 4. Lewiston, NY: Edwin Mellen, 1990.

Bossy, John. *Giordano Bruno and the Embassy Affair.* New Haven and London: Yale University Press, 1991.

Bouwsma, W. J. *Concordia Mundi: The Career and Thought of Guillaume Postel.* Cambridge, MA: Harvard University Press, 1957.

———. "Postel and the Significance of Renaissance Cabalism." *Journal of the Warburg and Courtauld Institutes* 18 (1954): 313–332.

Brann, Noel L. "The Shift from Mystical to Magical Theology in the Abbot Trithemius (1462–1516)." Ed. John R. Sonnenfeldt. *Studies in Medieval Culture* 11 (1977).

Breger, Herbert. "*Elias Artista*—a Precursor of the Messiah in Natural Science." In *Nineteen Eighty-Four: Science between Utopia and Dystopic*, ed. Everett Mendelsohn and Helga Nowotny, pp. 49–72. Sociology of the Sciences, vol. 8. New York: D. Reidel Publishing Company, 1984.

Brooks, David. "The Idea of the Decay of the World in the Old Testament, the Apocrypha, and the Pseudepigrapha." In *The Light of Nature*, ed. J. D. North and J. J. Roche, pp. 384–404. Dordrecht: Martinus Nijhoff Publishers, 1985.

Butler, E. M. *Ritual Magic.* Cambridge: Cambridge University Press, 1949.

Calder, I. R. F. "John Dee Studied as an English Neoplatonist." Dissertation, The Warburg Institute, London University, 1956.

Camden, Carroll. "The Wonderful Yeere." *Studies in Honor of De Witt T. Starnes*, ed. Thomas P. Harrison, pp. 163–179. Austin: University of Texas Press, 1967.

Capp, Bernard. *Astrology and the Popular Press: English Almanacs 1500–1800.* Ithaca, NY: Cornell University Press, 1979.

———. *The Fifth Monarchy Men: A Study in Seventeenth-Century English Millenarianism.* Totowa, NJ: Rowman and Littlefield, 1972.

Carlson, Marvin. *Places of Performance: The Semiotics of Theatre Architecture.* Ithaca, NY: Cornell University Press, 1989.

Chambers, E. K. *The Elizabethan Stage.* 4 vols. Oxford: The Clarendon Press, 1923.

Charbonneau-Lassay, Louis. *The Bestiary of Christ.* Ed. and trans. D. M. Dooling. New York: Parabola Books, 1991.

Charles, R. H. "The Book of Enoch." *Apocrypha and pseudepigrapha of the Old Testament.* 2 vols. Oxford: Oxford University Press, 1913.

Choisy, Maryse, ed. In collaboration with B. Grillot. *Anges, demons, et êtres intermédiares. Colloque, 13 et 14 janvier 1968.* Paris: Alliance Mondiale des Religions, Ed. Labergerie, 1969.

Christianson, Paul. *Reformers and Babylon: English Apocalyptic Visions from the Reformation to the Eve of the Civil War.* Toronto: University of Toronto Press, 1978.

Clucas, Stephen. "John Dee's *Liber Mysteriorum* and the *ars notoria*: Renaissance Magic and Medieval Theurgy." In *John Dee: Interdisciplinary Studies in Renaissance Thought,* ed. Stephen Clucas. Dordrecht: Kluwer Academic Publishers, forthcoming.

Clulee, Nicholas H. "Astrology, Magic, and Optics: Facets of John Dee's Early Natural Philosophy." *Renaissance Quarterly* 30 (1977): 632–680.

———. "At the Crossroads of Magic and Science: John Dee's Archemastrie." In *Occult and Scientific Mentalities in the Renaissance,* ed. Brian Vickers, pp. 57–71. New York: Cambridge University Press, 1984.

———. *John Dee's Natural Philosophy: Between Science and Religion.* New York: Routledge, 1988.

Cohn, Norman. *The Pursuit of the Millennium: Revolutionary Millenarians and Mystical Anarchists of the Middle Ages.* New York: Oxford University Press, 1970.

Copenhaver, Brian P. "Did Science Have a Renaissance?" *Isis* 93 (1992): 387–407.

———. "Lefèvre D'Etaples, Symphorien Champier, and the Secret Names of God." *Journal of the Warburg and Courtauld Institutes* 40 (1977): 189–211.

———. "Natural Magic, Hermetism and Occultism in Early Modern Science." In *Reappraisals of the Scientific Revolution,* eds. David Lindberg and Robert Westman, pp. 261–301. New York: Cambridge University Press, 1990.

Coudert, Allison. "Forgotten Ways of Knowing: the Kabbalah, Language, and Science in the Seventeenth Century." In *The Shapes of Knowledge from the Renaissance to the Enlightenment,* eds. D. R. Kelley and R. H. Popkin, pp. 83–99. Dordrecht: Kluwer Academic Publishers, 1991.

———. "Some Theories of a Natural Language from the Renaissance to the Seventeenth Century." In *Magia Naturalis und die Entstehung der Modernen Naturwissenschaften,* pp. 56–114. Wiesbaden: Franz Steiner Verlag, GMBH, 1978.

Cox-Rearick, Janet. *Dynasty and Destiny in Medici Art: Pontormo, Leo X, and the Two Cosimos.* Princeton, NJ: Princeton University Press, 1984.

Crombie, A. C. *Robert Grosseteste and the Origins of Experimental Science, 1100–1700.* Oxford: Clarendon Press, 1958.

Curtius, Ernst Robert. *European Literature and the Latin Middle Ages.* Trans. Willard R. Trask. Bollingen Series, vol. 36. Princeton, NJ: Princeton University Press, 1973.

Dales, Richard C. "The De-animation of the Heavens in the Middle Ages." *Journal of the History of Ideas* 41 (1980): 531–550.

Davidson, Gustav. *A Dictionary of Angels, Including the Fallen Angels*. New York: Macmillan, 1967.

Davies, Horton. *From Cranmer to Hooker 1534–1603*. Vol. 1 of *Worship and Theology in England*. Princeton, NJ: Princeton University Press, 1970.

Dear, Peter. *Discipline and Experience: The Mathematical Way in the Scientific Revolution*. Chicago: University of Chicago Press, 1995.

de Bellis, Carla. "Astri, Gemme e Arti Medico-magiche nello 'Speculum Lapidum' Di Camillo Leonardi." In *Il Mago, Il Cosmi, Il Teatro degli Astri: Saggi sulla Letteratura Esoterica del Rinascimento*, ed. Gianfranco Formichetti, intro. by Fabio Troncarelli, pp. 67–114. Rome: Bulzoni Editore, 1985.

Debus, Allen G. *The English Paracelsians (1965)*. New York: Franklin Watts, 1966.

———. "John Woodall, Paracelsian Surgeon." *Ambix* 10 (1962): 108–118.

———. "Mathematics and Nature in the Chemical Tracts of the Renaissance." *Ambix* 15 (1968): 1–28.

de Grazia, Margreta. "The Secularization of Language in the Seventeenth Century." *Journal of the History of Ideas* 41 (1980): 319–329.

Delatte, Armand. *La Catoptromancie grecque et ses dérivés*. Bibliothèque de la Faculté de Philosophie et Lettres de l'Université de Liège. Paris/Liège, 1932.

Dieckmann, Liselotte. "A Forgotten Alchemist, *Les Livres des figures hiéroglyphiques de Nicholas Flamel, écrivain*." *Festschrift für Bernhard Blume*, pp. 29–41. Göttingen: Vandenhoeck, 1967.

———. *Hieroglyphics: The History of a Literary Symbol*. Saint Louis, MO: Washington University Press, 1970.

———. "Renaissance Hieroglyphics." *Comparative Literature* 9 (1957): 308–321.

Dillon, John. *The Golden Chain: Studies in the Development of Platonism and Christianity*. Aldershot, Hampshire: Variorum, 1990.

Dobbs, Betty Jo Teeter. *Alchemical Death and Resurrection: The Significance of Alchemy in the Age of Newton*. Lecture sponsored by the Smithsonian Institution Libraries. Washington, DC: Smithsonian Institution Libraries, 1990.

———. *The Foundations of Newton's Alchemy*. Cambridge: Cambridge University Press, 1975.

———. "From the Secrecy of Alchemy to the Openness of Chemistry." In *Solomon's House Revisited: The Organization and Institutionalization of Science*, ed. Tore Fransmyr, pp. 75–94. Canton, MA: Science History Publications, 1990.

———. *The Janus Faces of Genius: The Role of Alchemy in Newton's Thought*. Cambridge: Cambridge University Press, 1991.

Easton, Stewart C. *Roger Bacon and His Search for a Universal Science*. Oxford: Basil Blackwell, 1952.

Eco, Umberto. *The Search for the Perfect Language.* Trans. James Fentress. Oxford: Blackwell, 1995.

Elert, Claes-Christian. "Andreas Kempe (1622–89) and the Languages Spoken in Paradise." *Historiographica Linguistica* 5 (1978): 221–226.

Eliade, Mircea, ed. *The Encyclopedia of Religion.* New York: Macmillan Publishing Company, 1987.

Epiney-Burgard, Georgette. "Jean Eck et le commentaire de la *Théologie Mystique* du Pseudo-Denys." *Bibliothèque D'humanisme et Renaissance* 34 (1972): 7–29.

Ernst, Germana. "From the Watery Trigon to the Fiery Trigon: Celestial Signs, Prophecies, and History." In *'Astrologi hallucinat': Stars and the End of the World in Luther's Time,* ed. Paola Zambelli, pp. 265–280. New York: Walter de Gruyter, 1986.

Evans, R. J. W. *Rudolf II and His World.* Oxford: Clarendon Press, 1973.

Faye, L. Kelly. *Prayer in Sixteenth-Century England.* Gainesville: University of Florida Press, 1966.

Feingold, Mordechai. *The Mathematicians' Apprenticeship: Science, Universities, and Society in England, 1560–1640.* Cambridge: Cambridge University Press, 1984.

———. "The Occult Tradition in the English Universities of the Renaissance: A Reassessment." *Occult and Scientific Mentalities in the Renaissance,* ed. Brian Vickers, pp. 73–94. New York: Cambridge University Press, 1984.

Feldman, W. M. *Rabbinical Mathematics and Astronomy.* New York: Sepher-Hermon, 1978.

Figurovski, N. A. "The Alchemist and Physician Arthur Dee (Artemii Ivanovich Dii): An Episode in the History of Chemistry and Medicine in Russia." *Ambix* 13 (1965): 35–51.

Findlen, Paula. "The Museum: Its Classical Etymology and Renaissance Genealogy." *Journal of the History of Collections* 1 (1989): 59–78.

Firpo, Luigi. "John Dee, scienziato, negromante e avventuriero." *Renascimento* 3 (1952): 25–84.

Firth, Katharine R. *The Apocalyptic Tradition in Reformation Britain 1530–1645.* Oxford: Oxford University Press, 1979.

Fontinoy, Charles. "Les Anges et les démons de L'Ancien Testament." In *Anges et Démons,* ed. Julien Ries with Henri Limet, pp. 177–134. Homo Religiosus 14. Louvain-le-Neuve: Centre D'Histoire Des Religions, 1989.

Fraade, Steven D. "Enoch." In *The Encyclopedia of Religion,* ed. Mircea Eliade, 5:116–118. New York: Macmillan Publishing Company, 1987.

Freiburgs, Gunar, ed. *Aspectus et affectus: Essays and Edition in Grosseteste and in Medieval Intellectual Life in Honor of Richard C. Dales.* Intro. Richard Southern. New York: AMS Press, 1993.

French, Peter. *John Dee: The World of an Elizabethan Magus.* New York: Routledge and Kegan Paul, 1972.

Froelich, Karlfried. "Pseudo-Dionysius and the Reformation of the Sixteenth Century." In *Pseudo-Dionysius: The Complete Works,* trans. Colm Luibheid, pp. 33–46. Classics of Western Spirituality. New York: Paulist Press, 1987.

Garrett, Cynthia. "The Rhetoric of Supplication: Prayer Theory in Seventeenth-Century England." *Renaissance Quarterly* 46 (1993): 328–357.

Geneva, Ann. *Astrology and the Seventeenth-Century Mind: William Lilly and the Language of the Stars.* Manchester: Manchester University Press, 1995.

Godwin, Joscelyn. *Music, Magic and Mysticism: A Sourcebook.* London: Arkana, 1987.

Goldberg, Benjamin. *The Mirror and Man.* Charlottesville: University of Virginia Press, 1985.

Gombrich, E. *Symbolic Images: Studies in the Art of the Renaissance.* New York: Phaedon, 1972.

Grafton, Anthony. *Defenders of the Text: The Traditions of Scholarship in an Age of Science.* Cambridge, MA: Harvard University Press, 1991.

———. *Historical Chronology.* Vol. 2 of *Joseph Scaliger: A Study in the History of Classical Scholarship.* Oxford-Warburg Studies. Oxford: Oxford University Press, 1994.

Graham, William A. *Beyond the Written Word: Oral Aspects of Scripture in the History of Religion.* Cambridge: Cambridge University Press, 1987.

Grant, Edward. *Planets, Stars, and Orbs: The Medieval Cosmos, 1200–1687* (1994). Cambridge: Cambridge University Press, 1996.

Graves, Robert, and Raphael Patai. *Hebrew Myths: The Book of Genesis.* New York: McGraw-Hill, 1963.

Greengrass, Mark, Michael Leslie, and Timothy Raylor, eds. *Samuel Hartlib and Universal Reformation: Studies in Intellectual Communication.* Cambridge: Cambridge University Press, 1994.

Grubb, Nancy. *Angels in Art.* New York: Artabras, 1995.

Hallyn, Fernand. *The Poetic Structure of the Universe* (1990). Trans. Donald M. Leslie. New York: Zone Books, 1993.

Hamilton, Alastair. *The Family of Love.* Cambridge: James Clarke and Company, 1981.

Hannaway, Owen. "Laboratory Design and the Aim of Science: Andreas Libavius versus Tycho Brahe." *Isis* 77 (1986): 585–610.

Harkness, Deborah E. "Managing an Experimental Household: The Dees of Mortlake." *Isis* 88 (1997): 242–262.

———. "The Scientific Reformation: John Dee and the Restitution of Nature." Ph.D. diss., University of California, Davis, 1994.

———. "Shows in the Showstone: A Theater of Alchemy and Apocalypse in the Angel Conversations of John Dee." *Renaissance Quarterly* 49 (1996): 707–737.

Harris, Roy, and Talbot J. Taylor. *Landmarks in Linguistic Thought: The Western Tradition from Socrates to Saussure.* New York: Routledge, 1989.

Heidt, William George. *Angelology of the Old Testament.* Washington, DC: Catholic University of America Press, 1949.

Heist, William W. *The Fifteen Signs before Doomsday.* East Lansing: Michigan State College Press, 1952.

Heninger, S. K., Jr. *The Cosmographical Glass: Renaissance Diagrams of the Universe*. San Marino, CA: The Huntington Library, 1977.

———. *Touches of Sweet Harmony. Pythagorean Cosmology and Renaissance Poetics*. San Marino, CA: The Huntington Library, 1974.

Hersey, G. L. *Pythagorean Palaces: Magic and Architecture in the Italian Renaissance*. Ithaca, NY: Cornell University Press, 1976.

Heym, Gerard. "Some Alchemical Picture Books." *Ambix* 1 (1937): 69–75.

Hill, Christopher. *The English Bible and the Seventeenth-Century Revolution*. London: Penguin, 1993.

Holstun, James. *A Rational Millennium: Puritan Utopias of Seventeenth-Century England and America*. Oxford: Oxford University Press, 1987.

Hopper, Vincent F. *Medieval Number Symbolism* (1938). New York: Cooper Square Publishers, 1969.

Horst, Irvin Buckwalter. *The Radical Brethren: Anabaptism and the English Reformation to 1558*. The Hague: B. de Graff, 1972.

Howell, A. C. "Res et Verba: Words and Things." *English Literary History* 13 (1946): 131–142.

Hunter, G. K. "Flatcaps and Bluecoats: Visual Signals on the Elizabethan Stage." *Essays and Studies*, n.s. 33 (1980): 16–47.

Hunter, Michael. "Alchemy, Magic, and Moralism in the Thought of Robert Boyle." *British Journal for the History of Science* 23 (1990): 387–410.

Idel, Moshe. *Kabbalah: New Perspectives*. New Haven and London: Yale University Press, 1988.

———. "The Magical and Neoplatonic Interpretation of the Kabbala in the Renaissance." Trans. Martelle Gavarin. In *Jewish Thought in the Sixteenth Century*, ed. Bernard Dov Cooperman, pp. 186–242. Cambridge, MA: Harvard University Press, 1983.

Innes, H. McLeod. *Fellows of Trinity College Cambridge*. Cambridge: Cambridge University Press, 1941.

Iversen, Erik. *The Myth of Egypt and Its Hieroglyphs in European Tradition*. Copenhagen: GecCad Publishers, 1961.

Jardine, Lisa. "Humanism and the Sixteenth-Century Cambridge Arts Course." *History of Education* 4 (1975): 16–31.

Jones, G. Lloyd. *The Discovery of Hebrew in Tudor England: A Third Language*. Manchester: Manchester University Press, 1983.

Jones, Richard Foster. *Ancients and Moderns: A Study of the Rise of the Scientific Movement in Seventeenth-Century England*. Berkeley: University of California Press, 1965.

Kagan, Richard L. *Lucrecia's Dreams: Politics and Prophecy in Sixteenth-Century Spain*. Berkeley: University of California Press, 1990.

Kieckhefer, Richard. *Magic in the Middle Ages*. New York: Cambridge University Press, 1990.

Klaassen, Walter. *Living at the End of the Ages: Apocalyptic Expectation in the Radical Reformation*. Lanham, MD: University Press of America, 1992.

Klossowski de Rola, Stanislas. *The Golden Game: Alchemical Engravings of the Seventeenth Century*. New York: G. Braziller, 1988.

Knoespel, Kenneth J. "The Narrative Matter of Mathematics: John Dee's Preface to the *Elements* of Euclid of Megara (1570). *Philological Quarterly* 66 (1987): 26–46.

Knowles, D. "The Influence of Pseudo-Dionysius on Western Mysticism." In *Christian Spirituality: Essays in Honour of Gordon Rupp*, ed. P. Brooks, pp. 79–94. London: SCM, 1975.

Knowlton, James. *Universal Language Schemes in England and France 1600–1800*. Toronto: University of Toronto Press, 1976.

Kristeller, Paul Oskar. *The Philosophy of Marsilio Ficino* (1943). Trans. Virginia Conant. Gloucester, MA: P. Smith, 1964.

Kruger, Steven F. *Dreaming in the Middle Ages*. Cambridge: Cambridge University Press, 1992.

Kuntz, Marion L. *Guillaume Postel, Prophet of the Restitution of All Things: His Life and Thought*. The Hague: Martinus Nijhoff Publishers, 1981.

———. "The Virgin of Venice and Concepts of the Millennium in Venice." In *The Politics of Gender in Early Modern Europe*, ed. Jean R. Brink, Allison P. Coudert, and Maryanne C. Horowitz, pp. 111–130. Sixteenth-Century Essays and Studies, vol. 12. Kirksville, MO: Sixteenth-Century Journal Publishers, 1989.

Laycock, Donald C. *The Complete Enochian Dictionary*. New York: S. Weiser, 1978.

Leader, Damian Riehl. *The University to 1546*. Vol. 1 of *A History of Cambridge University*. Cambridge: Cambridge University Press, 1988.

Lerner, Robert E. "The Black Death and Western European Eschatalogical Mentalities." *American Historical Review* 86 (1981): 533–552.

———. *The Powers of Prophecy: The Cedar of Lebanon Vision from the Mongol Onslaught to the Dawn of the Enlightenment*. Berkeley: University of California Press, 1983.

Levy, B. Barry. *Planets, Potions and Parchments: Scientific Hebraica from the Dead Sea Scrolls to the Eighteenth Century*. Montreal: McGill-Queen's University Press, 1990.

Lewis, C. S. *The Discarded Image: An Introduction to Medieval and Renaissance Literature*. Cambridge: Cambridge University Press, 1967.

Lindberg, David, trans. *Roger Bacon's Philosophy of Nature: A Critical Edition, with English Translation, Introduction, and Notes, of De multiplicatione specierum and De speculis comburentibus*. Oxford: Clarendon Press, 1983.

Long, Valentine. *The Angels in Religion and Art*. Paterson, NJ: St. Anthony Guild Press, 1970.

Lovejoy, Arthur O. *The Great Chain of Being: A Study of the History of an Idea*. New York: Harper Torchbooks/Academy Library—Harper and Row, 1960.

McConica, James, ed. *The History of the University of Oxford*. Oxford: Oxford University Press, 1986.

McEvoy, James. "The Metaphysics of Light in the Middle Ages." *Philosophical Studies* (Dublin) 26 (1979): 124–143.

———. *The Philosophy of Robert Grosseteste*. Oxford: Clarendon Press, 1982.

McKitterick, David. "Two Sixteenth-Century Catalogues of St. John's College Library." *Transactions of the Cambridge Bibliographical Society* 7 (1978): 135–155.

McKnight, Stephen A. *The Modern Age and the Recovery of Ancient Wisdom: A Reconsideration of Historical Consciousness 1450–1650*. Columbia: University of Missouri Press, 1991.

———. *Sacralizing the Secular: The Renaissance Origins of Modernity*. Baton Rouge and London: Louisiana State University Press, 1989.

———. "Science, the *prisca theologia*, and the Modern Epochal Consciousness." In *Science, Pseudo-science, and Utopianism in Early Modern Thought*, ed. Stephen McKnight, pp. 88–117. Columbia: University of Missouri Press, 1992.

Marsh, Christopher. *The Family of Love in English Society, 1550–1630*. Cambridge: Cambridge University Press, 1993.

Massetani, Guido. *La filosofia cabbalistica di Giovanni Pico della Mirandola*. Empoli: Tipografia di Edisso Traversari, 1897.

Milik, Jésef T., ed. *The Books of Enoch*. Oxford: Oxford University Press, 1976.

Miller, James. *The Cosmic Dance*. Toronto: University of Toronto Press, 1986.

Mills, David, ed. *Staging the Chester Cycle*. Leeds Texts and Monographs. Leeds: Moxon, 1985.

Moran, Bruce T. *The Alchemical World of the German Court: Occult Philosophy and Chemical Medicine in the Circle of Moritz of Hessen (1572–1632)*. Sudhoffs Archiv Zeitschrift für Wissenschaftsgeschichte, Beiheft 29. Stuttgart: Franz Steiner Verlag, 1991.

Moss, Jean Dietz. "The Family of Love and English Critics." *Sixteenth-Century Journal* 6 (1975): 33–52.

———. *"Godded with God": Hendrik Niclaes and His Family of Love*. Transactions of the American Philosophical Society, vol. 71. Philadelphia: The American Philosophical Society, 1981.

Nauert, Charles G. *Agrippa and the Crisis of Renaissance Thought*. Urbana: University of Illinois Press, 1965.

Nebelsick, Harold P. *The Renaissance, the Reformation and the Rise of Science*. Edinburgh: T.&T. Clark, 1992.

Nelson, Benjamin. "The Quest for Certitude and the Books of Scripture, Nature, and Conscience." In *The Nature of Scientific Discovery*, ed. Owen Gingerich, pp. 355–391. Washington, DC: Smithsonian Institution Press, 1977.

Newman, William R. *Gehennical Fire: The Lives of George Starkey, an American Alchemist in the Scientific Revolution*. Cambridge, MA: Harvard University Press, 1994.

Niccoli, Ottavia. *Prophecy and People in Renaissance Italy* (1987). Trans. Lydia G. Cochrane. Princeton, NJ: Princeton University Press, 1990.

North, J. D. "The Western Calendar: *Intolerabilis, horribilis, et derisibilis*; Four Centuries of Discontent." In *The Universal Frame*, pp. 39–78. London: The Hambledon Press, 1989.

Onians, Richard Broxton. *The Origins of European Thought about the Body, the Mind, the Soul, the World, Time, and Fate.* Cambridge: Cambridge University Press, 1951.

Orgel, Stephen. *The Illusion of Power: Political Theater in the English Renaissance.* Berkeley: University of California Press, 1975.

Owen, Alex. *The Darkened Room: Women, Power, and Spiritualism in Late Victorian England.* Philadelphia: University of Pennsylvania Press, 1990.

Pagel, Walter. *Paracelsus.* New York: S. Karger, 1958.

Patrides, C. A., and Joseph A. Wittreich, Jr., eds. *The Apocalypse in English Renaissance Thought and Literature.* Ithaca, NY: Cornell University Press, 1984.

Petersen, Rodney Lawrence. *Preaching in the Last Days: The Theme of "Two Witnesses" in the Sixteenth and Seventeenth Centuries.* Oxford: Oxford University Press, 1993.

Pitt, Joseph C. *Galileo, Human Knowledge, and the Book of Nature.* Dordrecht: Kluwer Academic Press, 1992.

Ponthot, Joseph. "L'Angélologie dans L'Apocalypse Johannique." In *Anges et Démons,* ed. Julien Ries with Henri Limet, pp. 301–312. Louvain-le-Neuve: Centre D'Histoire des Religions, 1989.

Popkin, Richard H. *The History of Scepticism from Erasmus to Descartes.* New York: Harper, 1968.

———, ed. *Millenarianism and Messianism in English Literature and Thought, 1650–1800.* Clark Library Lectures, 1981–1982. Leiden and New York: Brill, 1988.

Praz, Mario. *Studies in Seventeenth-Century Imagery.* 2 vols. Rome: Edizioni di storia e letteratura, 1964–1974.

Prest, John. *The Garden of Eden: The Botanic Garden and the Re-creation of Paradise.* New Haven and London: Yale University Press, 1981.

Quistorp, Heinrich. *Calvin's Doctrine of the Last Things.* Trans. Harold Knight. London: Lutterworth Press, 1958.

Rambaldi, Enrico I. "John Dee and Federico Commandino: An English and an Italian Interpretation of Euclid during the Renaissance." *Rivista di Storia della Filosofia* 44 (1989): 211–247.

Raspanti, Antonino. *Filosofia, teologia, religione: L'unita della visione in Giovanni Pico della Mirandola.* Palermo: Edi OFTES, 1991.

Read, John. *Prelude to Chemistry: An Outline of Alchemy.* Cambridge, MA: MIT Press, 1966.

Redgrove, H. Stanley. *Alchemy: Ancient and Modern.* New York: Barnes and Noble, 1922.

Reeds, J. "John Dee and the Magic Tables in the *Book of Soyga.*" In *John Dee: Interdisciplinary Studies in Renaissance Thought,* ed. Stephen Clucas. Dordrecht: Kluwer Academic Press, forthcoming.

———. "Solved: The Ciphers in Book III of Trithemius's *Steganographia.*" Forthcoming.

Reeves, Eileen. "Augustine and Galileo on Reading the Heavens." *Journal of the History of Ideas* 52 (1991): 563–579.

Reeves, Marjorie. *The Influence of Prophecy in the Later Middle Ages. A Study in Joachimism.* Oxford: Clarendon Press, 1969.

———. *Joachim of Fiore and the Prophetic Future.* New York: Harper Torchbooks—Harper and Row, 1977.

———. "Some Popular Prophesies from the Fourteenth to the Seventeenth Century." *Studies in Church History* 8 (1971): 107–134.

———, ed. *Prophetic Rome in the High Renaissance Period.* Oxford–Warburg Studies. Oxford: Oxford University Press, 1992.

Rekers, B. *Benito Arias Montana (1527–1598).* London: Warburg Institute, 1972.

Roques, René. *L'Univers Dionysien. Structure hiérarchique du monde selon le Pseudo-Denys.* Paris: Cerf, 1983.

Rowland, Christopher. *The Open Heaven. A Study of Apocalyptic in Judaism and Early Christianity* (1982). London: SPCK, 1985.

Salmon, Vivian. "Language-planning in Seventeenth Century England: Its Contexts and Aims." In *In Memory of J. R. Firth,* ed. C. E. Bazell et al., pp. 370–397. London: Longmans, 1966.

Scholem, Gershom. *Kabbalah* (1974). Library of Jewish Knowledge. Jerusalem: Keter Publishing House Jerusalem, 1977.

———. *Major Trends in Jewish Mysticism* (1941/1946). London: Thames and Hudson, 1955.

Secret, François. "Alchimie, palingénésie et métempsychose chez Guillaume Postel." *Chrysopoeia* 3 (1989): 3–60.

———. *Les Kabbalistes chrétiens de la Renaissance.* Paris: Dunod, 1964.

———. *Le Zôhar chez les Kabbalistes chrétiens de la Renaissance* (1958). Etudes Juives 10. Paris: Mouton, 1964.

Shapin, Stephen. " 'The Mind Is Its Own Place': Science and Solitude in Seventeenth-Century England." *Science in Context* 4 (1990): 191–218.

Sheppard, H. J. "Egg Symbolism in Alchemy." *Ambix* 6 (1958): 140–148.

Sherman, William H. "A Living Library: The Readings and Writings of John Dee." Dissertation, University of Cambridge, 1992.

———. *John Dee: The Politics of Reading and Writing in the Renaissance.* Amherst and Boston: University of Massachusetts Press, 1995.

Shumaker, Wayne. *John Dee's Astronomy.* Berkeley: California University Press, 1978.

———. "John Dee's Conversations with Angels." *Renaissance Curiosa,* pp. 15–52. Binghamton, NY: Center for Medieval and Early Renaissance Studies, 1982.

———. *The Occult Sciences in the Renaissance.* Los Angeles: University of California Press, 1972.

Simoncelli, Paolo. *La lingua di Adamo. Guillaume Postel tra accademici e fuoriusciti fiorentini.* Biblioteca della Rivista di Storia e Letteratura Religiosa Studi e Testi 7. Florence: L. Olschki, 1984.

Singer, Thomas C. "Hieroglyphs, Real Characters, and the Idea of Natural Language in English Seventeenth-Century Thought." *Journal of the History of Ideas* 50 (1989): 49–70.

Slaughter, M. M. *Universal Languages and Scientific Taxonomy in the Seventeenth Century*. New York: Cambridge University Press, 1982.

Smalley, Beryl. *The Study of the Bible in the Middle Ages* (1952). Oxford: Basil Blackwell, 1983.

Smith, Charlotte Fell. *John Dee (1527–1608)*. London: Constable and Company, 1909.

Smith, Pamela H. *The Business of Alchemy: Science and Culture in the Holy Roman Empire*. Princeton, NJ: Princeton University Press, 1994.

Smoller, Laura Ackerman. *History, Prophecy, and the Stars: The Christian Astrology of Pierre d'Ailly, 1350–1420*. Princeton, NJ: Princeton University Press, 1994.

Snelders, H. A. M. "Science in the Low Countries during the Sixteenth Century. A Survey." *Janus* 70 (1983): 213–227.

Southern, Richard. *Robert Grosseteste: The Growth of an English Mind in Medieval Europe*. Oxford: Clarendon Press, 1986.

Strong, Roy. *The Cult of Elizabeth*. New York: Thames and Hudson, 1981.

Tait, Hugh. "The Devil's Looking Glass." In *Horace Walpole: Writer, Politician, and Connoisseur*, ed. Warren H. Smith, pp. 195–212. New Haven and London: Yale University Press, 1967.

Taylor, E. G. R. *The Mathematical Practitioners of Tudor and Stuart England*. Cambridge: Cambridge University Press, 1967.

Tennant, F. R. *The Sources of the Doctrines of the Fall and Original Sin* (1903). Introd. by Mary Frances Thelen. New York: Schocken Books, 1968.

Thomas, Keith. *Religion and the Decline of Magic*. New York: Scribner's, 1971.

Thorndike, Lynn. *A History of Magic and Experimental Science*. 8 vols. New York: Columbia University Press, 1922–1948.

Tillyard, E. M. W. *The Elizabethan World Picture*. New York: Penguin, 1984.

Tomlinson, Gary. *Music in Renaissance Magic: Towards a Historiography of Others*. Chicago: University of Chicago Press, 1993.

Trevelyan, G. M. *Trinity College* (1943). Cambridge: Trinity College, 1972.

Tuveson, Ernest Lee. *Millennium and Utopia: A Study in the Background of the Idea of Progress*. Berkeley: University of California Press, 1949.

University of Louvain. *The University of Louvain 1425–1975*. Louvain: Louvain University Press, 1976.

Vandermeersch, Peter A. "The Reconstruction of the Liber Quintus Intitulatorum Universitatis Lovaniensis (1569–1616)." *Lias* 12 (1985): 1–80.

Van Dorsten, J. A. *The Radical Arts: First Decade of an Elizabethan Renaissance*. London: Oxford University Press, 1970.

Van Seters, John. "Elijah." In *Encylopedia of Religion*, ed. Mircea Eliade, 5: 91–93. New York: Macmillan Publishing Company, 1987.

Verbeke, Werner, Daniel Verhelst, and Andries Welkenhuysen, eds. *The Use and Abuse of Eschatology in the Middle Ages*. Louvain: Louvain University Press, 1988.

Vickers, Brian. "Analogy versus Identity: The Rejection of the Occult Symbolism, 1580–1680." *Occult and Scientific Mentalities in the Renaissance*, ed. Brian Vickers, pp. 95–163. Cambridge: Cambridge University Press, 1984.

Voet, Leon. *The Golden Compasses: The History of the House of Plantin-Moretus.* 2 vols. New York: Abner Schram, 1969.

Vondung, Klaus. "Millenarianism, Hermeticism, and the Search for a Universal Science." In *Science, Pseudo-science and Utopianism in Early Modern Thought*, ed. Stephen A. McKnight, pp. 118–140. Columbia: University of Missouri Press, 1992.

Walker, D. P. *Ancient Theology.* Ithaca, NY: Cornell University Press, 1972.

———. *Spiritual and Demonic Magic from Ficino to Campanella* (1958). Notre Dame, IN: University of Notre Dame Press, 1975.

Walton, Michael T. "John Dee's *Monas Hieroglyphica*: Geometrical Cabala." *Ambix* 33 (1976): 116–123.

Watson, Andrew G. "Thomas Allen of Oxford and His Manuscripts." *Medieval Scribes, Manuscripts and Libraries. Essays Presented to N. R. Ker*, pp. 279–316. London: Scolar, 1978.

Watts, Pauline Moffitt. "Prophecy and Discovery: On the Spiritual Origins of Christopher Columbus's 'Enterprise of the Indies.'" *American Historical Review* 90 (1985): 73–102.

Webster, Charles. *From Paracelsus to Newton: Magic and the Making of Modern Science.* Cambridge: Cambridge University Press, 1982.

Weil, G. E. *Elie Lévita: Humaniste et Massorète (1469–1549.* Leiden: E. J. Brill, 1963.

Weinstein, D. *Savonarola and Florence: Prophecy and Patriotism in the Renaissance.* Princeton, NJ: Princeton University Press, 1970.

Werblowsky, R. J. Zwi, and Geoffrey Wigoder, eds. *The Encyclopedia of the Jewish Religion.* New York: Adama Books, 1986.

Westman, Robert S. "Magical Reform and Astronomical Reform: The Yates Thesis Reconsidered. In *Hermeticism and the Scientific Revolution*, ed. Robert S. Westman and J. E. McGuire, pp. 1–91. Los Angeles: William A. Clark Memorial Library, 1977.

Whitby, Christopher. *John Dee's Actions with Spirits.* 2 vols. Dissertation, University of Birmingham, 1981. New York: Garland, 1988. (*See also under* Dee)

Wickham, Glynn. *Early English Stages 1300–1600.* 2 vols. New York: Columbia University Press, 1971.

Wigoder, Geoffrey, ed. *The Encylopedia of Judaism.* New York: Macmillan Publishing Company, 1989.

Wildiers, N. Max. *The Theologian and His Universe: Theology and Cosmology from the Middle Ages to the Present.* Trans. Paul Dunphy. New York: Seabury Press, 1982.

Williams, Ann, ed. *Prophecy and Millenarianism: Essays in Honour of Marjorie Reeves.* Essex: Longman, 1980.

Williams, Arnold. *The Common Expositor: An Account of the Commentaries on Genesis, 1527–1623.* Chapel Hill: University of North Carolina Press, 1948.

Wilson, Jean. *Entertainments for Elizabeth I.* Woodbridge, NJ: Brewer, 1980.

Wilson, Peter Lamborn. *Angels.* London: Thames and Hudson, 1980.

Wirszubski, Chaim. "Francesco Giorgi's Commentary on Giovanni Pico's Kabbalistic Theses." *Journal of the Warburg and Courtauld Institutes* 37 (1974): 145–156.

———. *Pico della Mirandola's Encounter with Jewish Mysticism.* Cambridge, MA: Harvard University Press, 1988.

Wittkower, Rudolf. "Hieroglyphics in the Early Renaissance." In *Developments in the Early Renaissance*, ed. Bernard S. Levy, pp. 58–97. Albany: SUNY Press, 1972.

Woolf, Rosemary. *The English Mystery Plays.* Berkeley: University of California Press, 1972.

Yates, Frances A. *The Art of Memory.* Chicago: University of Chicago Press, 1966.

———. *Astraea.* London: Routledge and Kegan Paul, 1985.

———. *Giordano Bruno and the Hermetic Tradition.* London: Routledge and Kegan Paul, 1964.

———. "Giordano Bruno's Conflict with Oxford." *Journal of the Warburg and Courtauld Institutes* 2 (1939): 227–242.

———. *The Occult Philosophy in the Elizabethan Age.* London: Routledge and Kegan Paul, 1979.

———. *Theatre of the World.* Chicago: University of Chicago Press, 1969.

Zambelli, Paola. "Magic and Radical Reformation in Agrippa of Nettesheim." *Journal of the Warburg and Courtauld Institutes* 39 (1976): 69–104.

Zika, Charles. "Reuchlin's *De Verbo Mirifico* and the Magic Debate of the Late Fifteenth Century." *Journal of the Warburg and Courtauld Institutes* 39 (1976): 104–138.

Index

Adam, 67, 68, 78, 79, 80, 81, 103, 126,
154, 158, 159, 160, 161, 162, 167,
169, 172, 177, 178, 179, 182, 193,
195, 196, 201, 207

Agnelli, Giovanni Baptista, 204 n. 31

Agrippa, Henry Cornelius, 35 n. 108, 38,
47, 49, 63, 85, 101, 102, 104, 105,
111, 118, 123, 127, 130, 160, 165,
167, 172, 176, 177, 180, 191, 192,
194, 202, 217, 225

Albertus Magnus, 20

Albumazar, 69

alchemy, 5, 20, 22, 28, 32, 33, 41, 49, 50,
51, 54, 55, 57, 63, 75, 77, 78, 88, 89,
90, 91, 96, 97, 148, 171, 180, 193,
194, 195–214, 215, 218, 221, 224–
225

medicine of God, 195, 203–204, 215

Aldaraia, see *Book of Soyga*

Aldrovandi, Ulisse, 62

Al-Kindi, 74, 75

Allen, Thomas, 20, 120

Ambrosius, Theseus, 167

angelology, 29, 31, 33–35, 37, 41, 43, 46–
51, 63, 85, 92, 103–130, 174

angel conversations

other interest in, 31 n. 91, 40, 103, 222–
223

properties of Dee's: books, 39–44: *Book
of Enoch*, 187, 192, 193, 194; *De
heptarchia mystica*, 29–30, 39, 41; 48
Claves Angelicae, 39, 41, 43, 44, 108
n. 183; *Liber Logaeth*, 39, 41, 42, 43,
45; *Liber scientiae auxilii et victoriae
terrestris*, 39, 43; *Mystery of Mysteries and the Holy of Holies*, 44; *Quartus liber mysteriorum*, 21; *Tabula
bonorum angelorum invocationes*, 39,

43; Holy Table: 29, 33–35, 37, 41,
219; lamine: 37–38; ring: 37–38; rod:
37; showstone: 23, 29, 30, 31, 32, 33,
37, 72, 77, 96, 99–100, 117, 118,
119, 127, 219, 221; wax seals: 33, 35,
41, 175, 180, 181, 184

theatrical conventions of Dee's, 205–214

angels

contacted by Dee: Anael: 50, 119, 127;
Bobogel: 186; Bornogo: 42; forty-nine
angelic Governors, 1a83, 184–185,
186, 187, 188, 189; Il, 35 n. 103,
146, 150; Ilemese, 51; Jublandace,
151; Gabriel, 37 n. 112, 47, 48, 49,
50, 82, 115, 176, 189, 203; Galuah,
115, 116, 177, 223; Levanael, 213;
Madimi, 115, 223; Metatron, 193;
Michael, 27, 37, 38, 44, 47, 48, 49,
82, 99, 115, 127, 208, 209; Murifri,
140, 141; Nalvage, 51, 157, 176, 188,
189; Och, 51; Raphael, 47, 48, 49,
115, 158, 178, 215, 216, 218; Salamian, 51; Sons and Daughters of
Light, 183–184, 187; of the thirty
Airs, 187, 189, 190, 191, 192; of the
twelve tribes of Israel, 187, 189, 190,
192; Uriel, 34, 35, 38, 44, 47, 49, 50,
115, 127, 140, 185, 208, 209

other efforts to contact, 31 n. 91, 40,
103, 222–223

religion of, 151–154, 156

Antwerp, 15, 47 n. 153, 85, 86, 129

apocalypticism, 3, 5, 16, 41, 44, 47, 48,
52, 59, 64, 68, 71, 104, 105, 107,
125, 126, 133–156, 162, 183, 189,
190, 195, 206, 207, 209, 210, 217,
222 (*see also* eschatology)

Aristophanes, 121